MW01470607

Immunohistochemistry

METHODS EXPRESS

The **METHODS EXPRESS** series

Series editor: B. David Hames

Faculty of Biological Sciences, University of Leeds, Leeds LS2 9JT, UK

Bioinformatics

Biosensors

Cell Imaging

DNA Microarrays

Expression Systems

Genomics

Immunohistochemistry

PCR

Protein Arrays

Proteomics

Whole Genome Amplification

Immunohistochemistry

METHODS EXPRESS

edited by S. Renshaw

Abcam Ltd, Cambridge, UK

Scion

First published 2007

A CIP catalogue record for this book is available from the British Library.

ISBN10: 1 904842 038 (paperback)
ISBN13: 978 1 904842 03 3
ISBN10: 1 904842 178 (hardback)
ISBN13: 978 1 904842 17 0

Scion Publishing Limited
Bloxham Mill, Barford Road, Bloxham, Oxfordshire OX15 4FF
www.scionpublishing.com

Important Note from the Publisher

The information contained within this book was obtained by Scion Publishing Limited from sources believed by us to be reliable. However, while every effort has been made to ensure its accuracy, no responsibility for loss or injury whatsoever occasioned to any person acting or refraining from action as a result of information contained herein can be accepted by the authors or publishers.

Typeset by Phoenix Photosetting, Chatham, Kent, UK
Printed by Cromwell Press, Trowbridge, UK

Cover image:
Immunofluorescent staining of human cells (U2O5) with an antibody towards alpha-tubulin, revealing a delicate network of alpha-tubulin (green FITC label) located exclusively in the cytoplasm. The nucleus is stained blue with a DAPI counterstain. Image © 2006 Abcam plc, and reproduced here with permission.

Contents

Contributors

Cuttle, Matthew University of Southampton School of Education, University Rd, Southampton SO17 1BJ, UK. E-mail: mfc206@soton.ac.uk

Jackson, Peter Department of Histopathology, Leeds General Infirmary, Leeds, LS1 3EX, UK. E-mail: pete.jackson@leedsth.nhs.uk

Jones, Ian Department of Biology and Biochemistry, University of Bath, Bath, BA2 7AY, UK. Current address: JT International SA, Chemin Rieu 14, 1211 Geneva 17, Switzerland. E-mail: ian.jones@jt-int.com

Mardle, Sarah Abcam plc, 332 Cambridge Science Park, Cambridge, CB4 0FW, UK. E-mail: sarah.mardle@abcam.com

Onley, Catherine Abcam plc, 332 Cambridge Science Park, Cambridge, CB4 0FW, UK. E-mail: catherine.onley@abcam.com

Powell, Janet Department of Anatomy, University of Cambridge, Downing St, Cambridge, CB2 3DY, UK. E-mail: jp247@cam.ac.uk

Renshaw, Simon Abcam plc, 332 Cambridge Science Park, Cambridge, CB4 0FW, UK. E-mail: simon.renshaw@abcam.com

Schenck, Emanuel Pfizer, IPC 330, Drug Safety Research & Development, PGRD, Ramsgate Rd, Sandwich, CT13 9NJ, UK. E-mail: emanuel.schenck@pfizer.com

Skepper, Jeremy Department of Anatomy, University of Cambridge, Downing St, Cambridge, CB2 3DY, UK. E-mail: jns1000@hermes.cam.ac.uk

Tannahill, David Wellcome Trust Sanger Institute, Wellcome Trust Genome Campus, Hinxton, Cambridge, CB10 1HH, UK. E-mail: dt2@sanger.ac.uk

Warford, Anthony Wellcome Trust Sanger Institute, Wellcome Trust Genome Campus, Hinxton, Cambridge, CB10 1HH, UK. E-mail: aw1@sanger.ac.uk

Westmacott, Adam Department of Biology and Biochemistry, University of Bath, Bath, BA2 7AY, UK. Current address: Abcam plc, 332 Cambridge Science Park, Cambridge, CB4 0FW, UK. E-mail: adam.westmacott@abcam.com

Wu, Jiahua Wellcome Trust Sanger Institute, Wellcome Trust Genome Campus, Hinxton, Cambridge, CB10 1HH, UK. E-mail: jerry.wu@sanger.ac.uk

Preface

Immunochemistry is an invaluable tool for the visualization of tissue antigens in diagnostic and biological research environments. The need to obtain accurate, reliable and reproducible results is paramount.

It is with this fundamental aim in mind that we have compiled *Immunohistochemistry: Methods Express*. We have achieved this by examining each aspect of immunochemistry in turn, with each chapter including detailed information regarding the subject matter in question. Each chapter is written by an expert in their field and includes protocols that are typically used in their own research. In addition, benefits and limitations of each approach are discussed within the chapters.

This book offers a wealth of knowledge to the novice immunochemist, who, from the outset, wishes to understand fully the theory and practice of immunochemical staining techniques and obtain reliable and reproducible data time and time again. For the experienced immunochemist, this book is a comprehensive reference guide to immunochemical staining techniques, allowing optimization of existing immunochemical staining protocols; and offering insight into advanced topics such as image capture, analysis and quantification.

Simon Renshaw

October 2006

Acknowledgements

On a personal note, I would like to thank the following people for their help, support and friendship at various stages along the way, both throughout this book and my career in general:

Abcam plc (employees of); Adam Westmacott; Addenbrooke's NHS Trust, Department of Histopathology (employees of); Alan Currie; Barbara Totty; Bruce Chalkley; Carol, Eddie, and Debbie Walton; Catherine Onley; Danielle Miller; David Hames; David Tannahill; David Williams; Eddie Powell; Emanuel Schenck; Ian Jones; Ian Scott; Jane Hoyle; Jeremy Skepper; Jim Neal; Jim Warwick; John Brown; Jonathan Milner; Jonathan Ray; Julian Beesley; Keith Miller; King's Mill NHS Trust, Department of Histopathology (employees of); Kwok Kee Chan; Lee, Nirvana, and Hannah Marles; Leslie Morris; Linda Cammish; Lisa Happerfield; Martin Coleman; Matt Cuttle; Matt Johnson; Michael Chipchase; Nick Coleman; Neil Hayward; Paul Trenam; Pete and Christine Burton; Peter Jackson; Queen's Medical Centre NHS Trust, Department of Histopathology (employees of); Robert Fincham; Sarah Mardle, Simon Brown, and Teresa Heath.

A special thank you goes to Robert Hughes, for giving me a lasting enthusiasm for histology and immunochemistry. Thank you for teaching me the importance of knowledge, self-progression and always being polite at a buffet!

A very special thank you goes to my parents, Gill and Arthur; and to my wife and daughter, Tracey and Lauren, for their continual patience and unconditional support. Thank you for always being there.

Abbreviations

ABC	avidin–biotin complex
AEC	3-amino-9-ethylcarbazole
AP	alkaline phosphatase
APES	aminopropyltriethoxysilane
BSA	bovine serum albumin
CCD	charge-coupled device
CD	cluster designation
CN	4-chloro-1-naphthol
COSHH	Control Of Substances Hazardous to Health
DAB	diaminobenzidine
ELISA	enzyme-linked immunosorbent assay
FITC	fluorescein isothiocyanate
HIER	heat-induced epitope retrieval
HRP	horseradish peroxidase
HSI	hue, saturation, and intensity
IMS	industrial methylated spirits
KLH	keyhole limpet hemocyanin
MAP	multiple antigenic peptide
NA	numerical aperture
NHS	normal human serum
OCT	optimal cutting temperature/tissue
PB	phosphate buffer
PBS	phosphate-buffered saline
PEG	polyethylene glycol
PSF	point spread function
RGB	red, green, and blue
RI	refractive index
scFv	single-chain Fv
SDS–PAGE	sodium dodecyl sulfate polyacrylamide gel electrophoresis
SQL	structured query language
TEM	transmission electron microscopy
TIRF	total internal reflectance
TMA	tissue microarray
UV	ultraviolet

Color section

Chapter 3. The selection of reporter labels

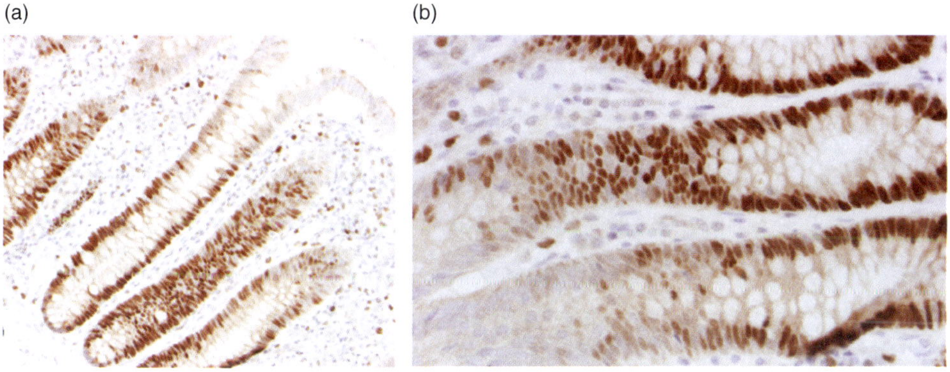

Figure 1. Immunochemical staining of human colon by a primary antibody raised against MCM 2 (see page 35).
The secondary antibody was conjugated to HRP and the chromogen used was DAB (see Chapter 3, Table 1). MCM 2 is a marker of cellular proliferation; hence, positive staining is predominantly demonstrated in the colonic crypts. Magnification: (*a*) ×20; (*b*) ×60.

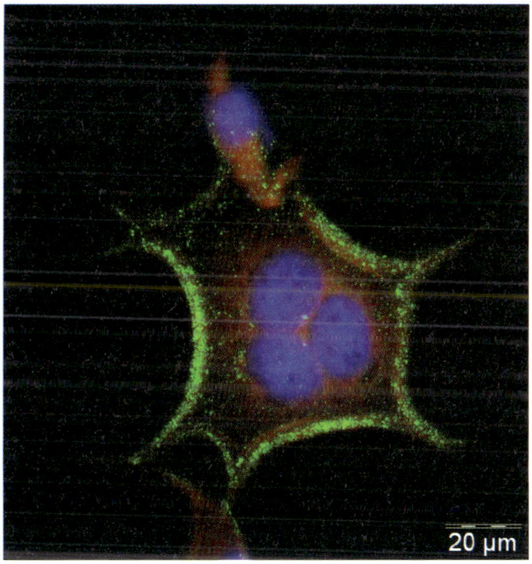

Figure 2. Immunofluorescent staining of human HeLa cells using a monoclonal antibody to pan-cadherin (see page 38).
The secondary antibody (green) was conjugated to Alexa Fluor 488. DAPI was used to stain the cell nuclei (blue). Alexa Fluor 594 conjugated to phalloidin was used to label F-actin (red).

Chapter 3. The selection of reporter labels

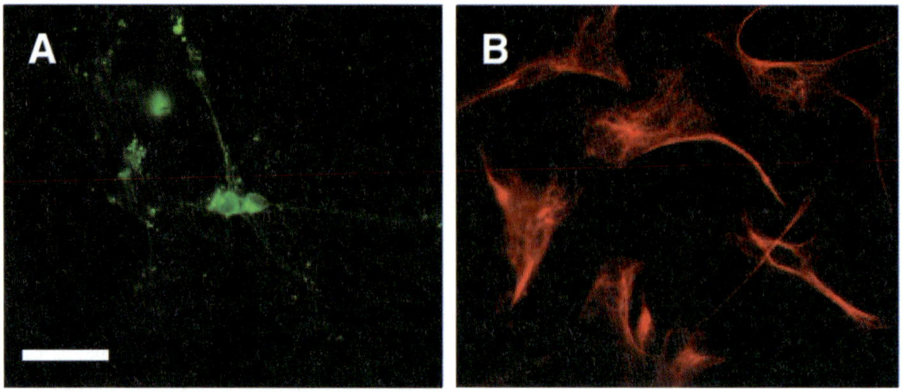

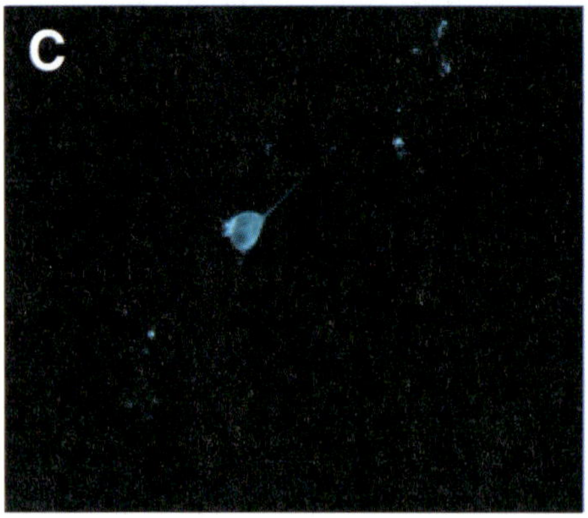

Figure 4. Immunofluorescence enables triple color images to be taken from the same sample (see page 43).
Here, a mixed population of cells can be distinguished comprising cultured neurons (stained with a primary antibody raised against a neuron-specific protein, detected with a secondary antibody conjugated to an FITC label, visualized at 518 nm) (*A*), glia (stained with a primary antibody raised against a glia-specific protein, detected with a secondary antibody conjugated to a Cy3 label, visualized at 570 nm) (*B*), and microglia (stained with a primary antibody raised against a microglia-specific protein, detected with secondary antibody conjugated to an Alexa Fluor 350 label, visualized at 442 nm) (*C*).

Chapter 5. Multiple immunochemical staining

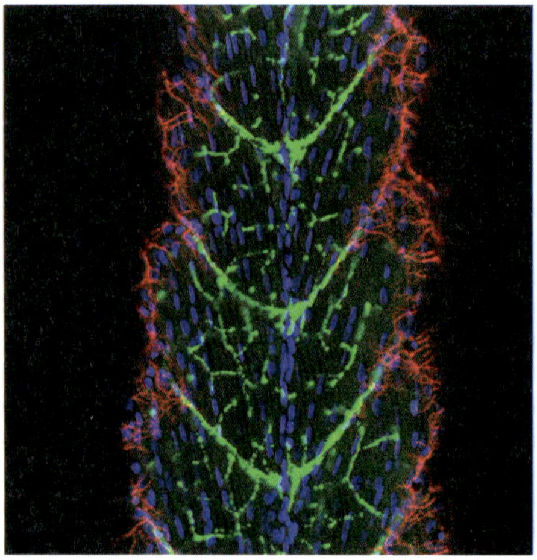

Figure 1. Confocal micrograph of an 8 day post-fertilization zebrafish trunk (see page 97).
The sample was immunohistochemically labeled with antibodies raised against a-bungarotoxin (green), tubulin (red) and Hu (neuronal cell marker; red). Nuclei were counterstained with TO-PRO3 (blue). Note the pronounced labeling of muscle cholinergic receptors by a-bungarotoxin in the somites. Image courtesy of Chris Church, University of Bath, UK.

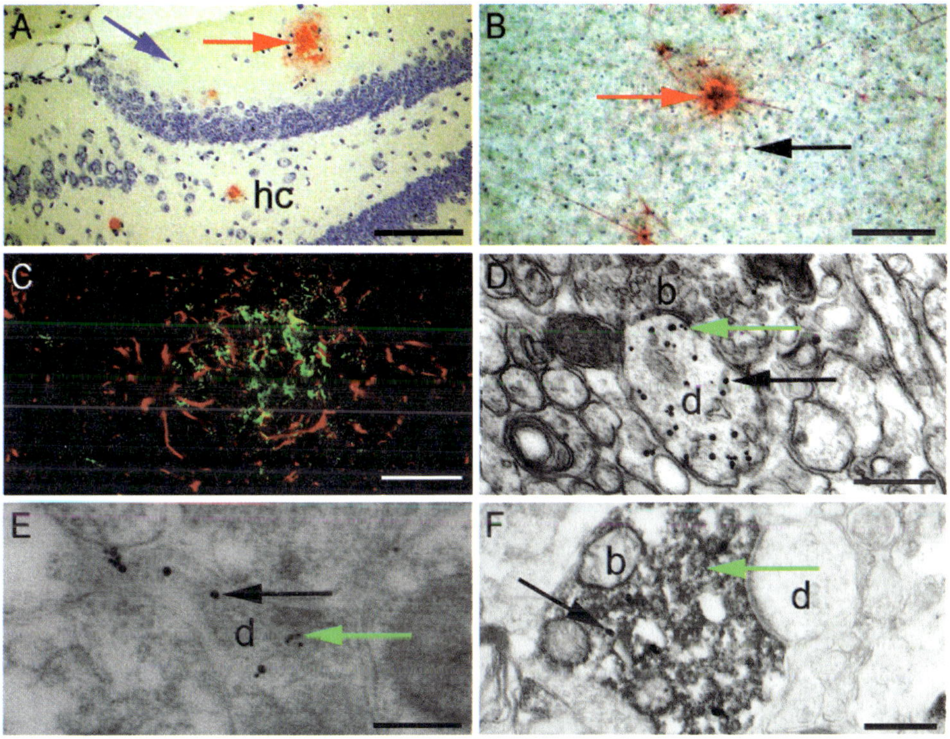

Figure 7. Examples of multiple staining at the light and electron microscope levels (see page 108). (*A*) Multiple immunoenzyme staining. Wax-embedded section through the mouse hippocampus dual labeled for glial fibrillary acidic protein (Vector VIP peroxidase substrate; red arrow) and β-amyloid (Vector Red alkaline phosphatase substrate; blue arrow) using *Protocol 1*. The section is counterstained with Nissl to reveal neuronal cell bodies (blue). (*B*) Immunoenzyme + immunogold dual staining for light microscopy. Wax-embedded section through the mouse hippocampus dual labeled for GFAP (Vector Red alkaline phosphatase substrate; red arrow) and α7 neuronal nicotinic acetylcholine receptors (nanogold-conjugated α bungarotoxin; black arrow) using *Protocol 4*. (*C*) Multiple immunofluorescence staining. Frozen section through the mouse hippocampus dual labeled for GFAP (Alexa Fluor 546-conjugated secondary antibody; red) and β-amyloid (Alexa Fluor 488-conjugated secondary antibody; green) using *Protocol 2*. (*D*) Double immunogold pre-embedding staining. Vibrating microtome section through the mouse midbrain dual labeled for vesicular glutamate transporters (large gold particles; black arrow) and α7 neuronal nicotinic acetylcholine receptors (small gold particles; green arrow) using *Protocol 3*. (*E*) Double immunogold post-embedding staining. Acrylate resin section through mouse midbrain dual labeled for tyrosine hydroxylase (large gold particles; black arrow) and β2 neuronal nicotinic acetylcholine receptor subunits (small gold particles; green arrow) using *Protocol 5*. (*F*) Immunoenzyme + immunogold dual staining for electron microscopy. Vibrating microtome section through the mouse midbrain dual labeled for α7 neuronal nicotinic acetylcholine receptors (gold particles; black arrow) and vesicular glutamate transporters (DAB peroxidase substrate; green arrow) using *Protocol 4*. b, Axonal bouton; d, dendrite; hc, hippocampus. Bars, 200 mm (*A*), 20 mm (*B*), 10 mm (*C*), and 200 nm (*D–F*).

Chapter 6. Confocal microscopy and immunochemistry

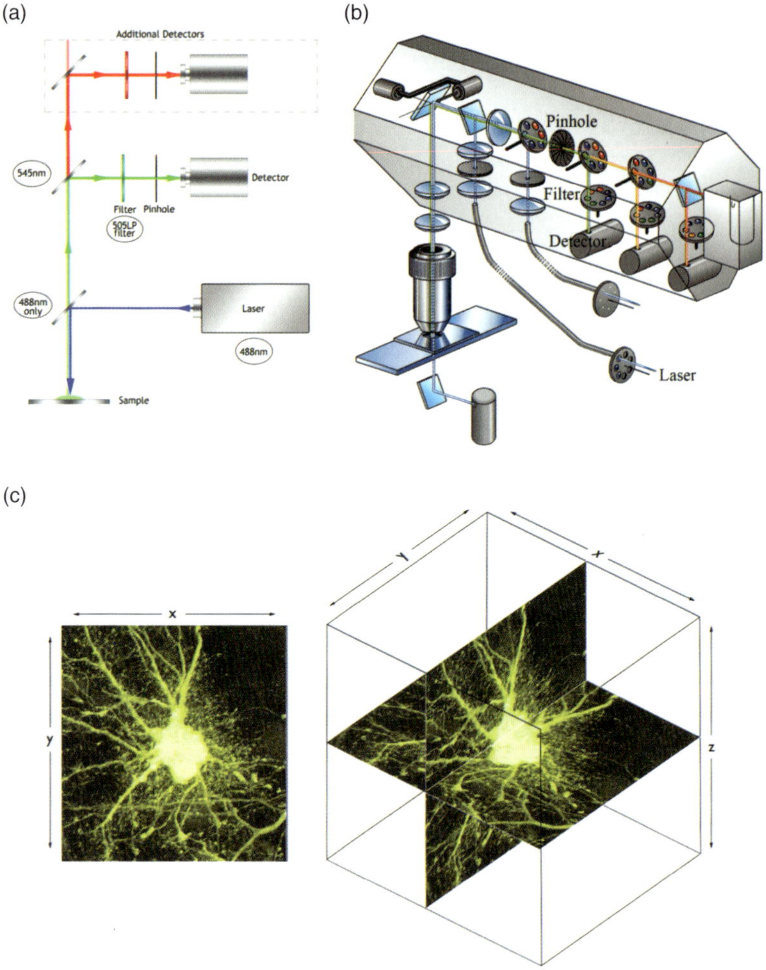

Figure 1. The inner workings of a confocal microscope needed to produce a three-dimensional reconstruction of the object (see page 130).
(*a*) The basic light path of a confocal microscope, with example filters for GFP (green fluorescent protein) imaging. The path starts with the laser light source (488 nm blue laser) and is reflected down to the sample with a dichroic mirror (488 nm only). Green fluorescence generated in the sample goes up and through the first dichroic mirror, is reflected off a second dichroic mirror (545 nm), through a filter (505 nm long pass (LP)), a pinhole and on to the photomultiplier tube detector. Additional detector(s) for longer-wavelength light can be added into the system beyond the second dichroic mirror, as shown. (*b*) A typical confocal light path inside a traditional, filter-based Leica microscope. The laser line, dichroic mirrors, filter wheels, and detectors can be seen, showing a light path similar to (*a*). In this example the pinhole is not directly in front of the detectors, but in front of the filter wheels. This configuration makes little practical difference to the system. (*c*) x, y, and z planes provide coordinates in a three-dimensional system, whereas two-dimensional systems (i.e. regular light microscopes) only demonstrate x and y planes.

Chapter 6. Confocal microscopy and immunochemistry

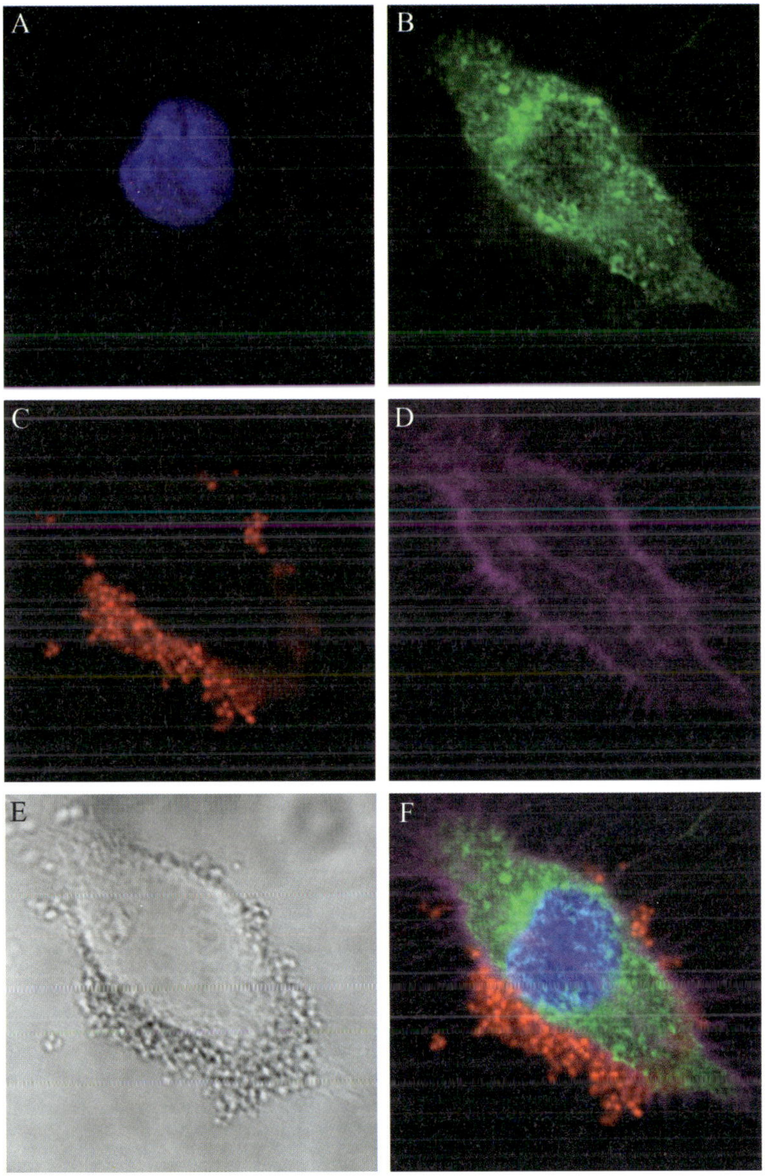

Figure 2. Confocal microscopy can be used to view multiple fluorophores in the same sample (see page 132).
In this example of staining in a single cell, the UV (364 nm) laser excites a blue nuclear stain (*a*), the blue (488 nm) laser excites a cytoplasmic dye, which fluoresces green (*b*), a bacterial marker excited by the green (543 nm) laser line fluoresces red (*c*) and phalloidin is used to mark F-actin and far-red fluorescence is generated by a red (633 nm) laser (*d*). A transmission image of the cell generated by the 633 nm laser line as longer-wavelength red light penetrates though the sample better than UV, blue, or green (*e*). The final plate shows the combined fluorescent image (*f*).

Chapter 6. Confocal microscopy and immunochemistry

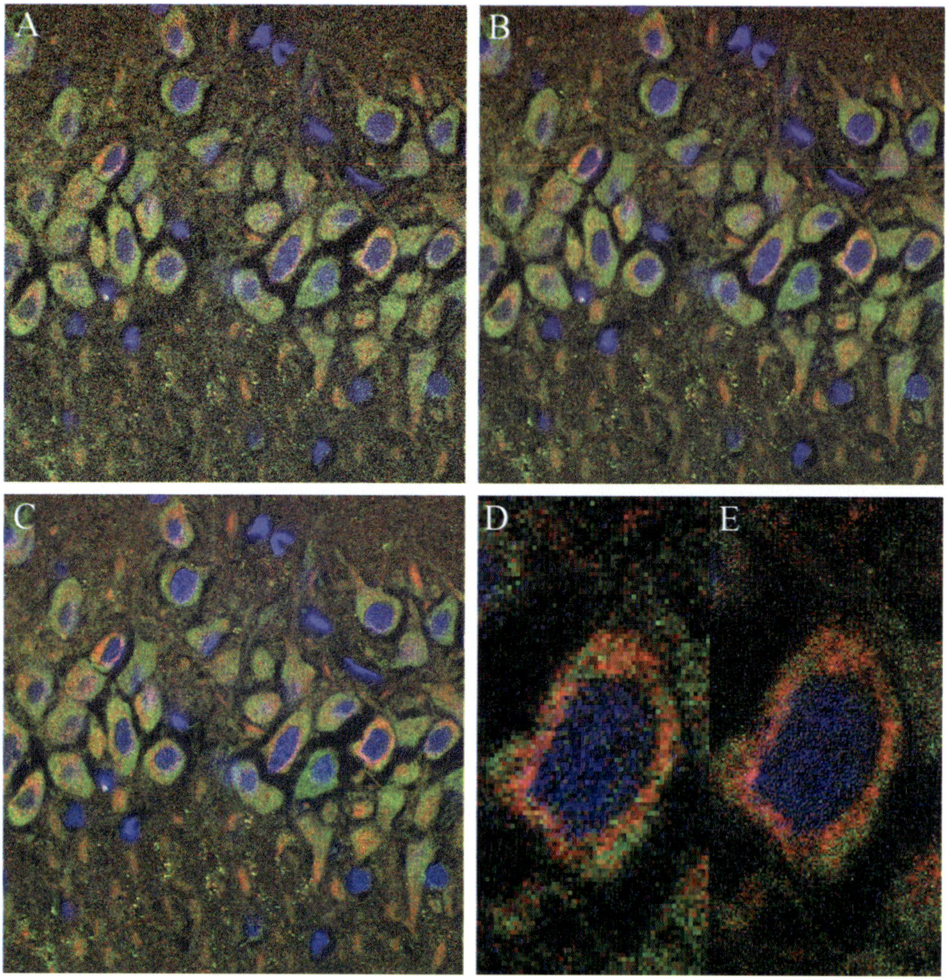

Figure 3. Image sampling and pixilation (see page 135).
(*A*) A three-color confocal image of the CA1 region of a hippocampal slice with blue nuclear staining, green cytoplasmic staining, and red protein staining. Image taken at 512 × 512 pixel resolution. (*B*) The same material sampled at 1024 × 1024 pixel resolution. Note the increase in image quality as a result of four neighboring pixels being averaged into a single image pixel. (*C*) The same material, collected at 512 × 512 pixel resolution, but as an average of four images. The images in (*B*) and (*C*) have the same apparent image quality, but the file size of (*C*) is four times smaller, so data handling and processing are significantly faster. Only if it is necessary to zoom in on a single cell does the difference between 512 × 512 imaging (*D*) and 1024 × 1024 imaging (*E*) become apparent, and under these conditions higher sampling rates are recommended.

Chapter 8. Image capture, analysis, and quantification

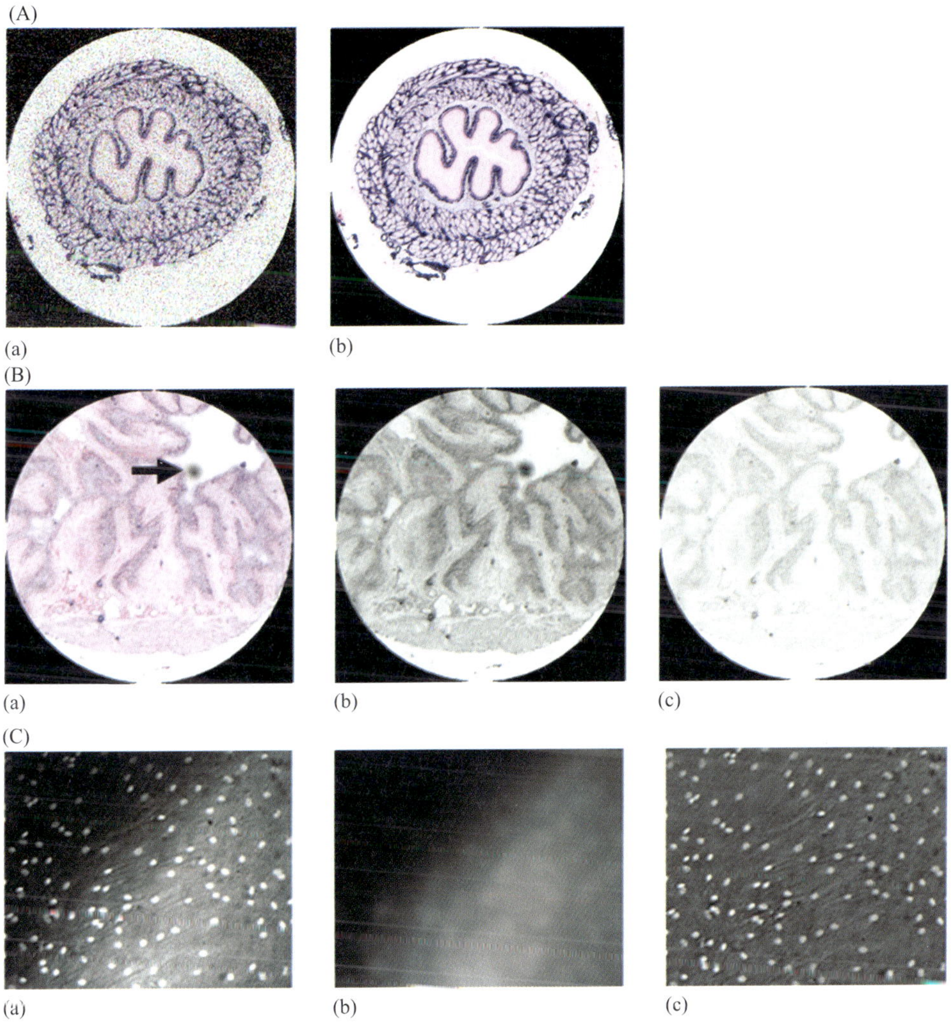

Figure 2. Correction of imaging defects (see page 185).
(*A*) Using a median filter. (*a*) A TMA core image displaying (artificial) background noise. (*b*) Median-filtered image of (a) showing noise removal. For the purposes of printed display, the noise in image (*a*) has been exaggerated. (*B*) Using a top-hat filter. (*a*) A TMA core image with background noise, which includes a large spot artifact (arrow). (*b*) Image showing color intensities of the image in (*a*). (*c*) Top-hat-filtered image showing the removal of the spot artifact. (*C*) Using a low-pass filter. (*a*) A field of nuclei with uneven illumination. (*b*) Estimated background image generated from a low-pass filter on (*a*). (*c*) Correction of uneven illumination by subtracting (*b*) from (*a*).

Chapter 8. Image capture, analysis, and quantification

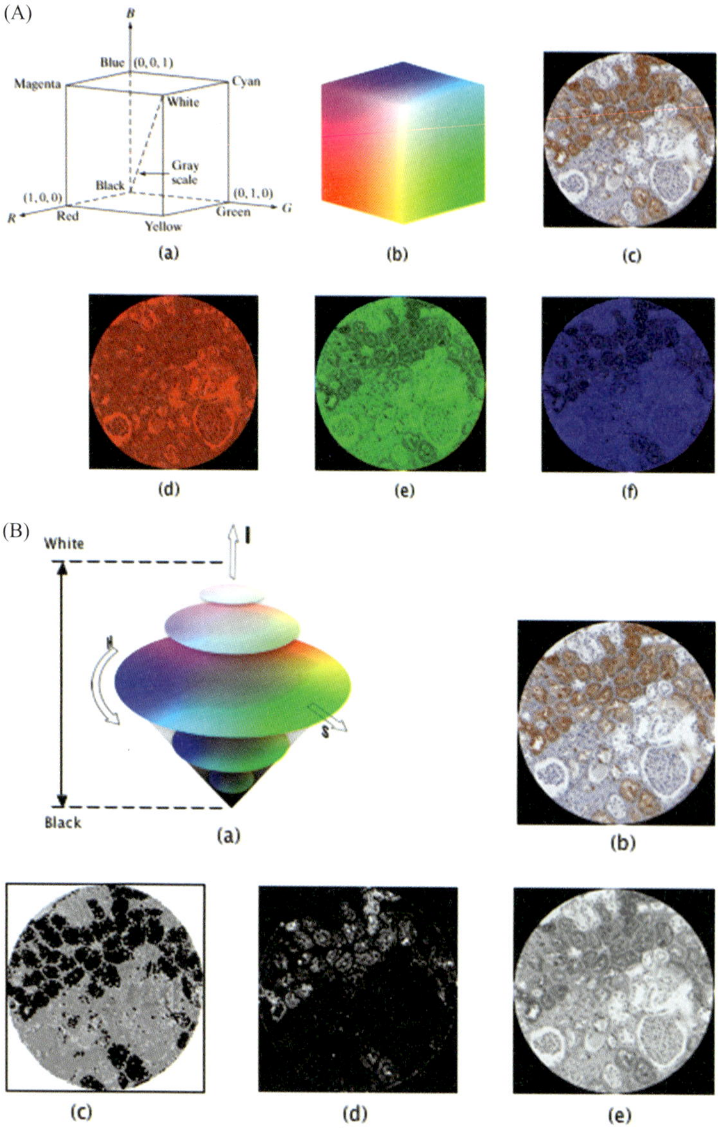

Figure 3. RGB and HSI color space (see page 186).
(*A*) RGB color space. (*a*) RGB color space can be geometrically represented in a 3D domain, where the coordinates of each point represent the values of the red, green and blue components, respectively. (*b*) RGB color cube. (*c*) Color image of a TMA core made up of a red component (*d*), a green component (*e*), and a blue component (*f*). (*B*) HSI color space. (*a*) Bi-conic representation of HSI color space. (*b*) Color image of a TMA core. (*c*) Hue component extract from (*b*). This is considered to be an angle between a reference line and a color point in HSI cone space. The range of hue values is from 0 to 360°. (*d*) Saturation component extracted from (*b*). This represents the radial distance from the cylinder centre; the nearer the point is to the center, the lighter the color. (*e*) Intensity component extracted from (*b*). This is the height in the *z*-axis direction. The axis of the cylinder describes the gray levels where zero intensity (minimum) is black and full (maximum) intensity is white.

Chapter 8. Image capture, analysis, and quantification

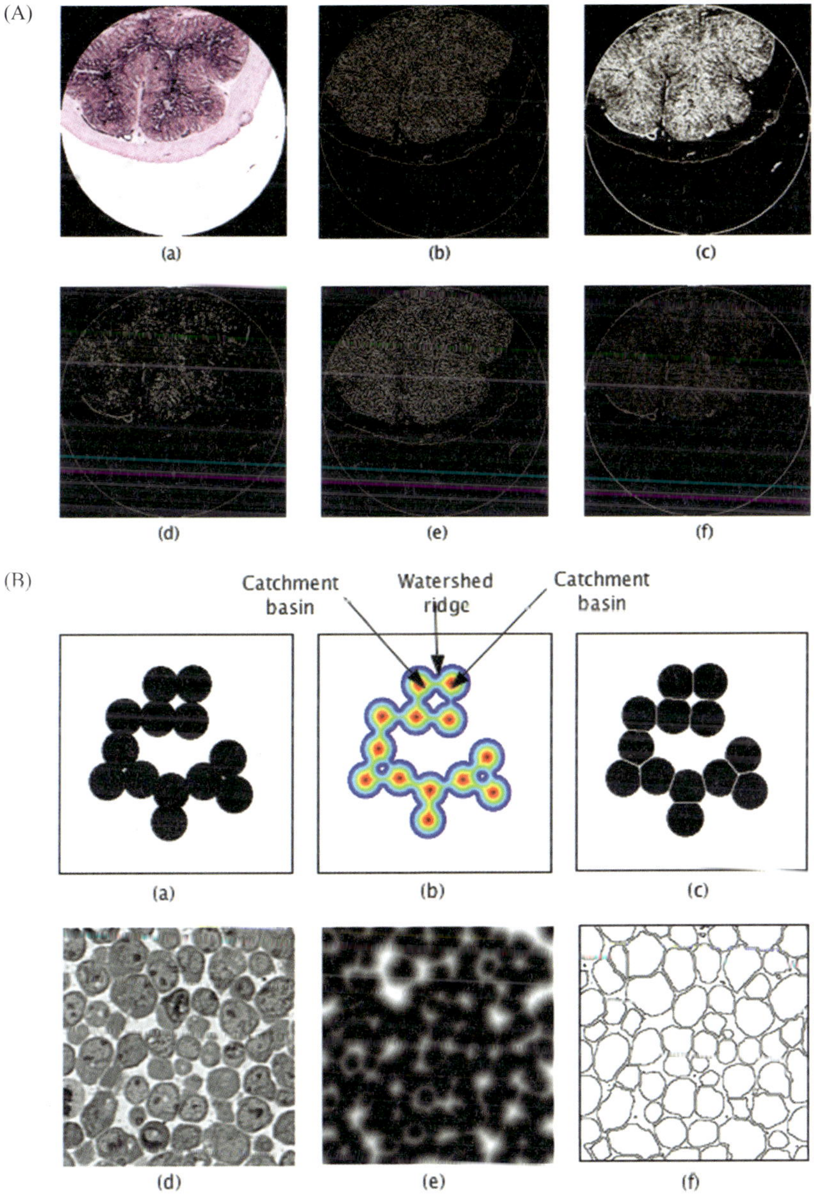

Figure 4. Image segmentation using edge detection and watershed segmentation (see page 190). (*A*) Use of edge detection filters. Original TMA core image (*a*) processed with Canny (*b*), Sobel (*c*), Prewitt (*d*), Lapacian (*e*) and Roberts (*f*) edge detection filters. (*B*) The principle of watershed segmentation. (*a*) A binary image of overlapping circles. (*b*) Euclidean distance transformation of the binary image in (*a*) with pixels color coded to show distance from boundary. A watershed is the ridge that divides areas drained by different river systems. A catchment basin is the geopgraphical area draining into a river or reservoir. (*c*) Result of watershed segmentation of (*a*). (*d*) Original image of a field of cells. (*e*) Watershed segmentation of the image in (*d*) using distance transformation. (*f*) Outline of the watershed-segmented cells in (*d*).

Chapter 8. Image capture, analysis, and quantification

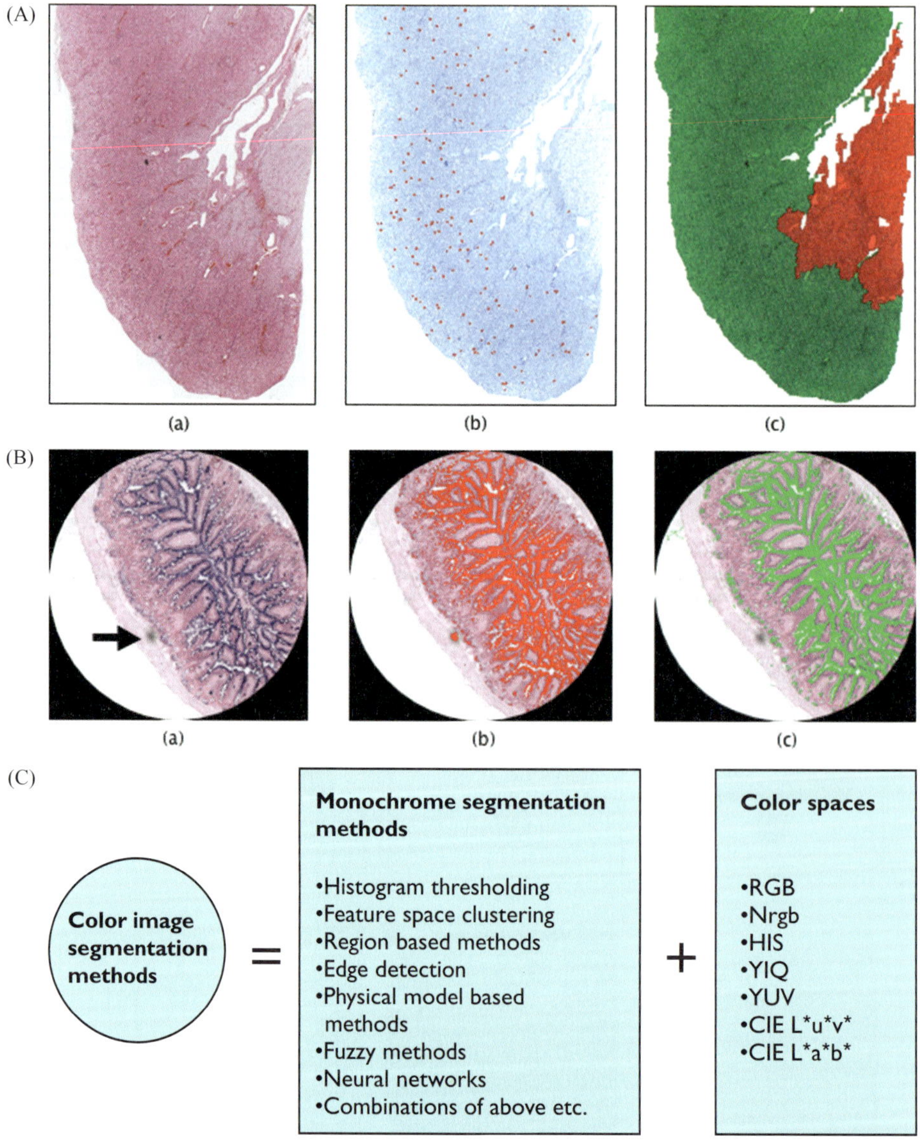

Figure 5. Image segmentation using texture-based segmentation, morphological operation, and color segmentation (see page 192).
(*A*) Texture-based segmentation. (*a*) Original image of a mouse kidney section. (*b*) Identification of kidney tissue from background (blue) and glomeruli identification (red). (*c*). Segmentation of cortex (green) and medulla (red). Note that these images were processed by CELLENGER software (images courtesy of The Wellcome Trust Sanger Institute/Definiens AG). (*B*) Removal of contamination by a morphological operation. (*a*) A TMA core image with a contamination spot (arrow). Note that the color blue represents the staining. (*b*) The mask image of segmentation for the staining without any morphological operations (red). (*c*) The mask image of segmentation for the staining after several morphological operations (green). Note that the contamination has been removed successfully in this mask (i.e. the spot is not green). (*C*) Common color segmentation approaches.

Chapter 8. Image capture, analysis, and quantification

(a)

(b)

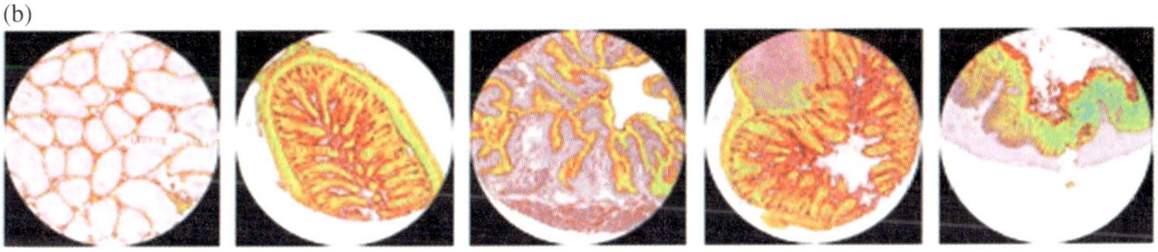

(c)

Tissue factor	0.785	0.600	0.836	0.777	0.498
Stain factor	0.156	0.898	0.433	0.718	0.556
Density factor	0.202	0.269	0.248	0.242	0.333

Figure 6. Example of staining intensity quantification **(see page 196).**
(*a*) Five different TMA cores with positive alkaline phosphatase staining (blue). (*b*) Segmentation of the blue staining in (*a*). Here, a pseudo-color intensity map has been derived from only the regions of blue staining and is shown laid over the TMA core; moderate/strong intensity is seen as yellow. (*c*) Quantitative analysis of the images in (*a*) by area analysis. The tissue factor is the tissue area/core area and represents the amount of tissue in the core. The stain factor is stain area/tissue area and represents the area of tissue stained. The density factor is the average stain intensity per pixel (normalized, where 0 is minimum and 1 is maximum intensity) and is a measure of signal strength.

Chapter 8. Image capture, analysis, and quantification

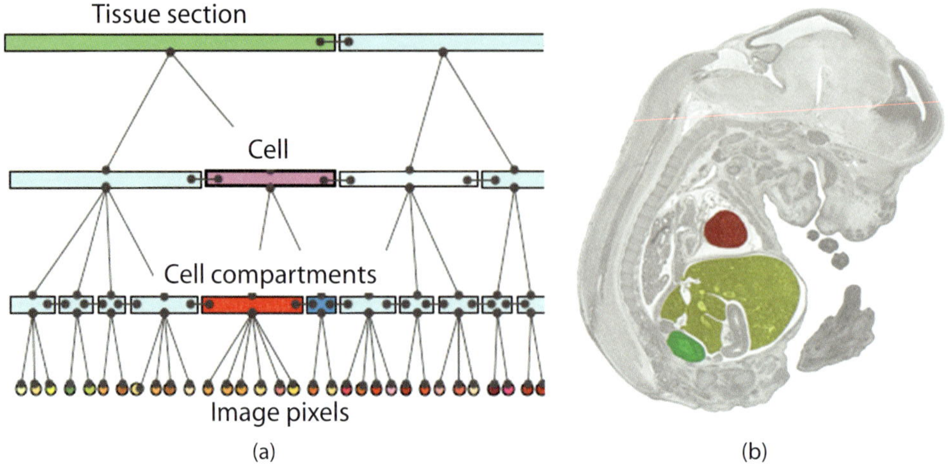

Figure 7. Object-orientated image analysis (see page 197).
In this example, an E14.5 mouse embryo section was processed using a rule set in the CELLENGER package to identify uniquely different embryonic tissues. (*a*) Example of the image object hierarchies. (*b*) Result of image processing showing the separation of different tissue regions, e.g. heart (red); liver (yellow) and kidney (green) (images courtesy of The Wellcome Trust Sanger Institute/Definiens AG).

CHAPTER 1
Introduction

Simon Renshaw

Immunochemistry is the identification of a certain antigen in a histological tissue section or cytological preparation by an antibody specific to that antigen. The localization of the primary antibody (and therefore the target antigen) is then visualized using the appropriate microscope by a gold, enzymatic or fluorescently labeled detection system. There are numerous techniques available, depending on the nature of the specimen and on the degree of sensitivity required.

Before moving any further, it is important to consider the difference between immunohistochemistry and immunocytochemistry. Immunohistochemistry refers specifically to histological tissue sections (frozen or paraffin embedded) and immunocytochemistry refers specifically to cytological preparations, whether in the form of a conventional smear or a monolayer. Generally (and for the majority of this book), the terms immunohistochemistry and immunocytochemistry can be considered essentially the same, but they are often used without regard to their true meaning. This can cause problems, especially when a third party is involved, such as technical assistance from an antibody supplier. The very nature of the specimen dictates whether or not certain immunochemical strategies can be employed, and confusing the two can lead to inappropriate steps of an immunochemical staining protocol being included or omitted. For the purpose of clarification throughout the book, where no distinction needs to be made between immunohistochemistry and immunocytochemistry, the terms 'immunochemical staining' or 'immunochemistry' will be used.

The term 'immunofluorescence' is also often used without regard for its true meaning. Most people who speak of immunofluorescence mean immunocytochemistry using an antibody labeled with a fluorochrome. In fact, 'immunofluorescence' is a broad term referring to any immunochemical staining technique involving a fluorescent label. However, where no distinction needs to be made, the term 'immunofluorescent staining' or indeed 'immunofluorescence' will be used.

Immunohistochemistry: *Methods Express* (S. Renshaw, ed.)

Immunochemistry has proved an invaluable tool in diagnostics as well as in a biological research environment. Pathologists can usually make a diagnosis using tissue sections stained with traditional tinctorial dyes, such as the time-honored hematoxylin and eosin, but immunochemistry is often used to enforce the diagnosis further. For instance, the classification of lymphomas has been greatly aided by using an appropriate anti-CD marker panel. The need for accurate, reliable, and reproducible results in diagnostic immunochemistry has largely pioneered today's plethora of high-quality antibody reagents and sensitive detection systems that are available today. Automation in the immunochemical staining laboratory has provided a reliable means of high-throughput immunochemical staining, saving valuable time and human resources that can be used elsewhere to provide a better diagnostic service or accelerate the rate of biological research.

The aim of this book is to take the reader in a logical order through the various theoretical and practical aspects of immunochemical staining. The concept of antibodies and antibody production can be found in Chapter 2, moving on to detection systems in Chapter 3. The concepts of both chapters are then combined to examine common immunochemical staining techniques in Chapter 4. Once basic immunochemical staining techniques have been established, the book then moves on to specialized immunochemical staining techniques. Chapter 5 deals with multiple immunochemical staining, followed by Chapter 6, which is concerned with confocal microscopy in conjunction with multiple immunochemical staining. Chapter 7 then addresses ultrastructural electron microscopy when combined with immunogold staining. In a biological research or teaching environment, careful analysis and quantification of immunochemical staining image results is paramount, a topic covered in Chapter 8. It is very important to emphasize at this early stage of the book that the overall aim of an immunochemical staining experiment is to provide accurate, specific, and reproducible results, the technicalities of which are explored in Chapter 9. Finally, Chapter 10 looks at automated immunochemical staining systems that are now commonplace in most laboratories.

This book offers a wealth of knowledge to the novice immunochemist, who, from the outset, wishes to understand fully the theory and practice of immunochemical staining techniques and to obtain reliable and reproducible data time and time again. For the experienced immunochemist, this book is a comprehensive reference guide to immunochemical staining techniques, allowing optimization of existing immunochemical staining protocols, and offering insight into advanced topics such as image capture, analysis, and quantification.

CHAPTER 2
Antibodies for immunochemistry

Catherine Onley

1. INTRODUCTION

The quality and suitability of antibody reagents are fundamental for the success and reproducibility of immunochemical studies. This chapter focuses on the selection and application of antibody reagents for practical users of immunochemistry. It starts by providing an overview of antibody biochemistry for researchers who are new to this area and then summarizes key considerations for antibody production and commercial sourcing of antibodies for immunochemistry. Finally, it provides recommended protocols for peptide antigen design.

Antibodies are a central component of the adaptive immune response. The role of antibodies *in vivo* is to detect foreign molecules, bind these molecules specifically, and then trigger immune responses such as activation of complement and phagocytosis. In order to mount an appropriate defense, the antibody must have exquisite specificity for the foreign molecule. It is this ability of antibodies to specifically recognize a myriad of different molecules that makes them indispensable tools for biochemical research and particularly for the localization of biological molecules in cells and tissues, forming the basis of immunochemical staining.

1.1 Antibodies are a family of glycoproteins with a central role in the adaptive immune response

Antibodies are members of the immunoglobulin (Ig) family of proteins and, in mammals, comprise five main classes or isotypes: IgA, IgD, IgE, IgG, and IgM (see *Table 1*). The majority of research antibody reagents are IgG or IgM.

The fundamental antibody structure is a Y-shaped molecule formed from two identical heavy chains and two identical light chains, stabilized

Immunohistochemistry: *Methods Express* (S. Renshaw, ed.)

by disulfide bonds (see *Fig. 1*). Both heavy and light chains contain variable and constant domains. The variable (V) domains contain the majority of amino acid sequence variation and mediate binding of the antibody to the target molecule or antigen (**anti**body **gen**erator). The amino acid sequence of the heavy chain constant (C) domain defines the isotype of the antibody. There are only two types of light chain, kappa and lambda. These associate with the N-terminal region of the heavy chains independent of antibody isotype; hence, the molecular formula of IgG can be either $\gamma_2\kappa_2$ or $\gamma_2\lambda_2$ and the formula of pentameric IgM is either $(\mu_2\kappa_2)_5$ or $(\mu_2\lambda_2)_5$ (see *Table 1*). The light chains contain one constant and one variable domain.

Table 1. Properties of human immunoglobulin classes.

	IgG	IgM	IgA	IgD	IgE
Heavy chain	γ	μ	α	δ	ε
Light chain	κ/λ	κ/λ	κ/λ	κ/λ	κ/λ
Subunit composition	$\gamma_2\kappa_2/\gamma_2\lambda_2$	$(\mu_2\kappa_2)_5/(\mu_2\lambda_2)_5$	$(\alpha_2\lambda_2)_n/(\alpha_2\kappa_2)_n$*	$\delta_2\kappa_2/\delta_2\lambda_2$	$\varepsilon_2\kappa_2/\varepsilon_2\lambda_2$
Molecular weight (kDa)	150	970	160/320	175	190
Antigen binding sites (valency)	2	10 (5)	2, 4 or 6	2	2
Concentration in human sera	8–16 mg/ml	0.5–2 mg/ml	1.4–4 mg/ml	3–40 µg/ml	0.017–0.45 µg/ml

*Where n = 1, 2, or 3.
Adapted from (1).

The prototypic immunoglobulin, IgG, is abundant in most tissues and in the circulation. IgG is a monomer composed of one copy of the basic Y-shaped antibody subunit and has a molecular weight of around 150 kDa. The heavy chain of IgG consists of three constant domains (C_H1, C_H2, and C_H3) and one variable domain (V_H), as shown in *Fig. 1*. There is a high degree of flexibility in the IgG molecule, which contributes to antigen binding. IgG molecules have a wide range of affinities to antigen, and high-affinity IgG can be generated. Along with ease of production and purification, this contributes to the widespread use of this class of antibody in immunoassays.

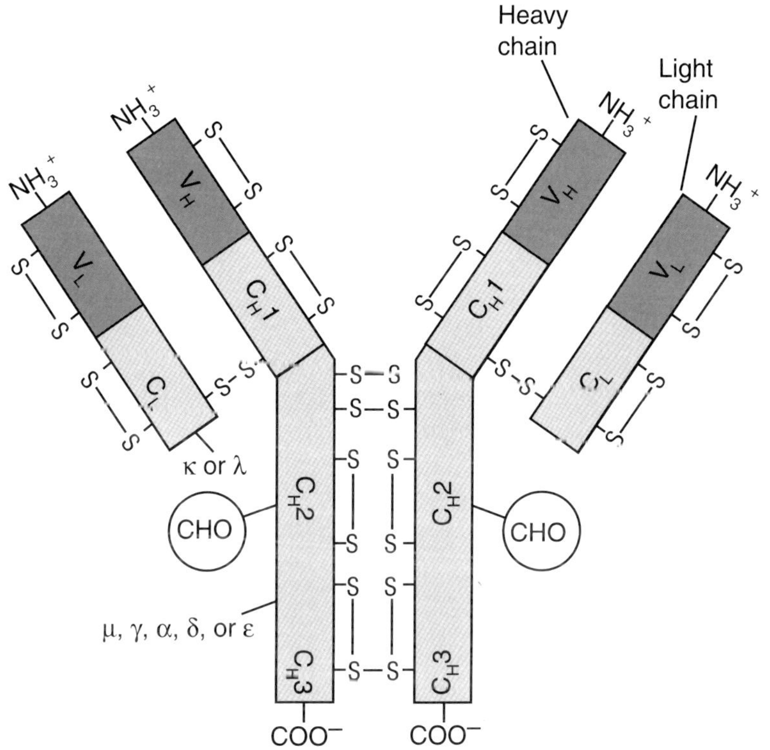

Figure 1. Schematic of the structure of IgG.
The basic Y-shaped antibody molecule contains four polypeptide chains and is stabilized by disulfide bonds (as indicated). The number and position of interchain disulfide bonds is dependent upon the antibody subclass. The heavy (H) and light (L) chains contain several immunoglobulin domains referred to as the constant (C) and variable (V) domains. The constant domains have a conserved amino acid sequence and dictate the antibody isotype. The sequence variation required to specifically recognize a diverse range of antigens is contained within the variable domains. Antibodies are glycoproteins and the carbohydrate moieties, which are present in the C_H2 domains, contribute to antibody structure and effector function.

IgG is synthesized during the secondary immune response and complexes of IgG and antigen activate the classical complement pathway. The effector functions of IgG are also mediated through binding of IgG complexes to a diverse group of cell-surface receptors, referred to as Fcγ receptors. Binding to Fcγ receptors is mediated by the Fc region of IgG. A full description of the complexity of effector functions mediated by IgG complexes is beyond the scope of this book but excellent discussion and further references can be found in immunology texts such as (1).

IgM, referred to historically as macroglobulin, has a more complex multimeric structure than IgG. The most common form has a molecular weight of around 900 kDa and contains five Y-shaped subunits linked by a single 15 kDa J-chain. Hence, IgM has a theoretical maximum valency of 10. The hinge region of IgG is replaced with an additional constant domain, which reduces the flexibility of IgM relative to IgG. When in complex with larger molecules, the valency of IgM can be reduced to 5, which is attributed to steric hindrance due to this lack of flexibility in the antibody molecule. IgM is largely confined to the bloodstream and appears early in the immune response. IgM tends to have a low affinity for antigen; however, the high valency of IgM leads to high-avidity binding in situations where multiple epitopes are presented. *Table 1* gives additional details of the structural and biological features of immunoglobulin isotypes.

1.2 Isotypic, allotypic, and idiotypic variations contribute to the diverse repertoire of antibody structures

In addition to the five main antibody isotypes, differences in the amino acid sequence and disulfide bonding patterns of the heavy chain give rise to antibody subclasses. Subclasses of IgG are found in a number of species; in mouse, for example, the IgG subclasses are designated $IgG_{1,}$ $IgG_{2,}$ IgG_{2a}, and IgG3. The subclass of mouse monoclonal antibodies should be known for antibodies used in immunochemical staining, as this will impact on experimental design, including selection of the correct subclass of isotype control. Commercial nonspecific isotype control antibodies are widely available. Antibody subclasses should also be considered during the development of antibody purification strategies, as they confer differential binding to protein A and G (see *Table 2*).

Also of relevance to end-users of antibodies in immunochemical staining studies are the advantages of defined IgG subclasses for multilabeling studies where subclass-specific secondary antibodies can be used to good effect to expand the range of targets that can be labeled in parallel (see Chapter 5, section 1.4.2).

Unlike antibody isotypes, which are expressed at the same time in the host, antibody allotypes are specific to the individual and are due to the presence of allelic forms of immunoglobulin molecules. Antibody allotypes

Table 2. The affinity of antibodies for proteins A, G, and L is dependent on host species, antibody isotype, and subclass

	Protein L[a]	Protein A[b]	Protein G[c]
Human Ig			
IgG	++	+++	+++
IgM	++	+	–
IgA	++	+	–
IgE	++	+	–
IgD	++	+	–
Fab	++	+	+
F(ab′)2	++	+	+
κ light chains	++	–	–
scFv	++	+	–
Mouse Ig			
IgG1	++	+	++
IgG2a	++	++	++
IgG2b	++	++	++
IgG3	++	+	++
IgM	++	+	–
IgA	++	++	+
Polyclonals			
Mouse	++	++	++
Rat	++	+	++
Rabbit	+	++	+++
Sheep	–	++	++
Goat	–	+	++
Bovine	–	+	++
Porcine	++	++	++
Chicken IgY	++	–	–

[a]Protein L birds primarily to κ light chains and λVI chains.
[b]Protein A birds to γFc and VHIII variable domains.
[c]Protein G binds to γFc and cγl chains
+++ = very strong binding, ++ = strong binding, + = moderate binding, – = no binding

Reproduced from Harlow and Lane (1988) with permission from Cold Spring Harbor Laboratory Press.

do not influence antibodies used in standard immunochemical staining techniques.

The unique tailoring of antibody molecules to specifically recognize a single epitope on the antigen is due to the generation of antibody idiotypes. Here, protein sequence changes occur in the hypervariable regions, which form the antigen-binding sites. The diverse repertoire of antibodies specific for a vast array of antigens arises solely from sequence variation within these regions.

1.3 The structures of antibodies are optimized for antigen binding and effector functions

The immunoglobulin superfamily of proteins has a modular domain structure containing one or more conserved globular domains, the immunoglobulin domain. This family of proteins includes immunoglobulins (antibodies) and a large number of cell-surface receptors, including Fc receptors and major histocompatibility complex class I and II molecules. The immunoglobulin domain consists of around 110 amino acids and has a characteristic β-pleated sheet structure stabilized by intrachain disulfide bonds. The immunoglobulin domains in antibodies correspond to the constant and variable domains discussed above. Loops (β-turns) protrude from the stable central scaffold of the domain. In the antibody variable domains of both the heavy and light chains, these loops (5–10 amino acid residues in length) contain hypervariable sequences that form the antigen-binding site. Three discrete loops (or hypervariable regions) are contributed to the antigen-binding site by each of the heavy and light chains. These loops confer antigen recognition and are also referred to as complementarity-determining regions.

The structural immunoglobulin domain is optimized for stability in the extracellular environment and molecular recognition. Antibodies are particularly stable molecules and are able to withstand a wide range of conditions without losing their biological function. Intermolecular disulfide bonds between the heavy and light chains and *N*-linked glycosylation also contribute to antibody stability. Treatment of IgG with reducing agents, such a 2-mercaptoethanol or dithiothreitol, cleaves the intermolecular disulfide bridges and separates the heavy and light chains, giving the classical 50 and 25 kDa fragments (see *Fig. 2*). These antibody fragments can be resolved using sodium dodecyl sulfate-polyacrylamide gel electrophoresis (SDS-PAGE) under reducing conditions. SDS-PAGE is often used to check for antibody integrity and to verify the lack of contaminating sera proteins in purified antibody samples.

Antibody glycosylation is an active area of research, particularly in therapeutics, where the glycosylation pattern of the antibody may be influenced by *in vitro* production techniques. Glycosylation contributes to the structure and stability of the antibody molecule, having an indirect role in antigen recognition. Glycosylation is also likely to be involved in antibody effector functions. In IgG, the C_H2 domain contains a single conserved *N*-glycosylation site (Asn-297) although further *N*-glycosylations may also be present. Carbohydrate moieties in the Fab region of IgG can prevent antigen binding and are responsible for the

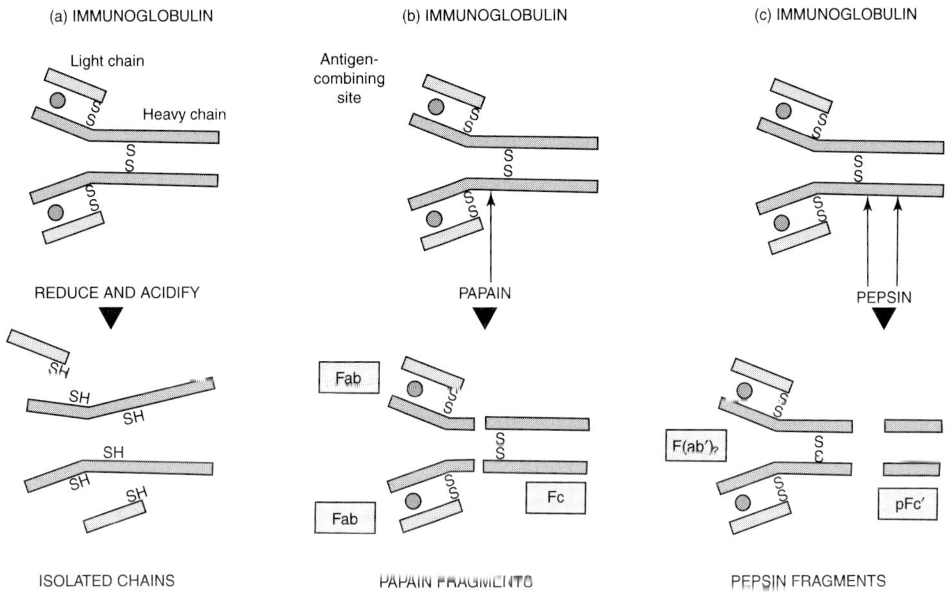

Figure 2. IgG is susceptible to reduction and enzymatic cleavage.
(*a*) The disulfide bonds that link the heavy and light chain of the fundamental antibody structure can be reduced to give fragments of 50 and 25 kDa. (*b*) The antibody unit is also susceptible to enzymatic cleavage. Papain digestion forms two fragments, Fab and Fc. Fab fragments retain antigen-binding properties. (*c*) Pepsin cleaves the C terminus of the hinge region giving a divalent $F(ab')_2$. Redrawn from *Roitt's Essential Immunology* with permission from Blackwell Publishing.

proportion (up to 30% of rabbit and human IgGs) of nonprecipitating, asymmetric, monovalent IgG.

Antibodies can be separated by site-specific proteolysis into three types of antibody fragment, monovalent Fab (fragment having the antigen-binding site), divalent $F(ab)'_2$ and Fc (fragment that crystallizes) (see *Fig. 2*). To generate Fab and Fc fragments from IgG, the extended proline-rich hinge region is cleaved using the enzyme papain, followed by a purification step. The lack of Fc receptor binding and effector functions in Fab fragments can be advantageous. However, being monovalent, the strength of binding of Fab fragments to the target antigen can be lower than the parent molecule due to loss of avidity effects. To counter this, pepsin can be used to cleave the C terminus of the interchain disulfide, generating $F(ab')_2$. These contain two antigen-binding sites giving a co-operative binding effect and an avidity for the target more comparable to the intact antibody. Commercial kits are available for the preparation and purification of Fab and $F(ab')_2$. These usually give reliable and reproducible results.

1.4 Multiple factors contribute to the antibody–antigen interaction, including antibody affinity, avidity, and specificity

The complementary and stereospecific binding of antibody to antigen was demonstrated by Karl Landsteiner in the 1930s through studies of antibody recognition of small molecules (2). Although small molecules and haptens have been used extensively to investigate the molecular basis of the antibody–antigen interaction, they are often not immunogenic in their own right and require conjugation to an immunogenic carrier molecule. As most antibodies used in immunochemical staining techniques are against protein targets, this section will focus on the properties of protein and peptide antigens (for a review, see 3).

Antigen recognition is conferred by the three complementarity-determining regions contributed to the antigen-binding site by V_H and V_L. These form a continuous hypervariable surface consisting of 110–130 amino acid residues referred to as the antigen-binding site (or paratope). A number of intermolecular forces contribute to antigen recognition and stabilization of the binding of antibody to antigen, including van der Waals interactions, hydrogen bond networks, and electrostatic and hydrophobic interactions. X-ray crystal structures of a number of antibodies (or antibody fragments) in complex with antigen have helped to elucidate how the strength and specificity of the antibody–antigen interaction depends on combinations of these interactions, which lead to a precise orientation of the two molecules (reviewed in section 2.2). Water molecules also contribute to complementarity by introducing additional hydrogen bonding opportunities and van der Waals interactions within the binding site. Aromatic residues, which are capable of multiple types of interaction, are prevalent in the antigen-binding site and in antigenic sites, with tyrosine in particular being over-represented.

The antibody-binding site (or epitope) on protein antigens can be formed from continuous or discontinuous amino acid sequences. The majority of epitopes for antibodies raised against native proteins are discontinuous, being clustered in the protein's three-dimensional conformation (tertiary structure). Hence, the conformation of the protein is a significant determinant of antibody recognition when the antibody is raised against a native or recombinant protein. It is often cited that antibodies that recognize the native conformation of the target protein work well in immunochemical applications. However, this observation should be viewed with caution, as many of the traditional methods employed in immunochemistry such as fixation, tissue processing, and antigen retrieval may affect the target protein's tertiary structure. These changes in the tertiary protein structure can be due to chemical modification of amino acids, such as oxidation of cysteine and methionine, reduction of disulfide bonds, and modification of primary amines on lysine residues or the N terminus by cross-linking fixatives such as formaldehyde. In addition, gross changes in protein structure can be caused by denaturation, fixation, proteolytic cleavage, changes in pH, and dehydration (see Chapter 4, section 2.1).

Antibodies raised against peptide immunogens that mimic continuous epitopes on the target protein may detect both folded and denatured protein, depending on the success of the immunogen selection and the individual immune response of the host. Therefore, antibodies need to be evaluated in a particular technique on an individual basis and an antibody showing exquisite specificity and affinity under some experimental conditions may be useless under different conditions, or in another application. It is difficult, and maybe impossible, to generalize about the likely success of an antibody in immunochemical staining techniques.

2. METHODS AND APPROACHES

2.1 The relationship between antibody affinity, avidity, and specificity

The first consideration for selecting antibodies for immunochemical staining is whether the antibody will recognize the antigen sufficiently to allow clear detection, often after fixation, and in the context of the cellular environment. The second is to minimize cross-reactivity with other proteins. The question of whether the antibody will recognize the antigen after sample preparation is often best addressed by trial and error and by employing a range of fixatives, tissue preparation (including antigen retrieval), and staining techniques (see Chapter 9, sections 2.1, 2.4, and 2.5, respectively). A working knowledge of the biochemical basis of antibody affinity, avidity, and specificity can help to interpret experimental results.

2.1.1 Affinity

Antibody affinity can be defined as the strength of binding of one epitope to a monovalent antibody, such as an Fab fragment, and is expressed in terms of the association constant, K_a. In equilibria where the associated form of the antigen and antibody is favored, the antibody is referred to as being of higher affinity. K_a is expressed in units of M^{-1} (1/moles per liter) and the higher the value, the higher the affinity. For instance, in immunochemical techniques, antibodies with an affinity value (K_a) of 10^6 M^{-1} are likely to give a weaker signal than those with higher K_a values such as 10^8 M^{-1}. K_a values may be as high as 10^{12} M^{-1} for the highest-affinity antibodies. Whilst the affinity of monoclonal antibodies can be determined, the affinity of polyclonals cannot, as polyclonal sera contains a mixture of antibodies of different affinities. The binding of an antibody to an antigen is an equilibrium, and most antibody–antigen interactions reach equilibrium in less than 1 h at room temperature. This will be slower at 4°C, hence the use of longer antibody incubations (usually overnight) at lower temperatures. The antibody–antigen interactions are reversible and the dynamic loss of binding and rebinding to the same or another site is often referred to as 'breathing'.

2.1.2 Avidity

Antibody avidity also contributes to the success of immunochemical staining. Avidity is a measure of the contribution of multivalent interactions to the stability of the antigen–antibody complex. This can occur where an antigen contains multiple copies of an epitope or where there is a high local concentration of the protein of interest. The inherent flexibility in antibody structure allows recognition of multiple antigens in an array of conformations. For IgG, this is mediated primarily by flexibility in the hinge region between the Fc and the Fab, which allows lateral and rotational movement of the Fab. Once a multivalent complex is formed, the rate of dissociation, K_d $(1/K_a)$, of the antibody from a single epitope is the same. However, the antibody is tethered by other interactions, which facilitate the reassociation with the original or neighboring epitopes and hence the observed rate of dissociation is less. Antigens displaying multiple epitopes can be cross-linked by antibodies to form stable complexes. Formation of these complexes in immunochemical staining can enhance the experimental results, by leading to signal amplification. If either the antibody or the antigen is in excess, this may limit the formation of these complexes. This phenomenon is called prozone and is observed infrequently in immunochemical staining experiments, but is one reason why more is not always better for immunochemical applications and why titration of the primary antibody is essential during the optimization of an immunoassay. Multivalent interactions of antigen with primary antibodies also enhance the possibility of multivalent interactions of secondary antibody. Hence, lower-affinity antibodies, which may not be appropriate for techniques such as immunoprecipitation, may give good results in immunochemistry due to high local concentrations of antigen, which contribute to high-avidity binding. However, low-affinity cross-reactivity with proteins other than the target can also be a particular problem in immunochemical techniques (see Chapter 9, section 2.6.2). Unwanted cross-reactivity and background staining is also often heavily influenced by primary antibody concentration and other protocol issues (for full details concerning removal of background staining, please refer to Chapter 9, sections 2.8 and 3).

2.1.3 Specificity

Specificity is harder to determine in immunochemical staining when compared with Western blotting or immunoprecipitation. In immunochemistry, subcellular localization is often limited to nuclear or cytoplasmic staining. Therefore, particular care must be taken to incorporate appropriate controls including negative-control antibodies, control tissues or cell samples, detection antibody controls, and multiple distinct antibodies to the target protein; this is essential to validate observations (controls for immunochemistry are discussed in more detail in Chapter 9, section 2.6). As always, the best test of specificity in immunochemical staining techniques is the use of knockout tissues or cells. However, even interpretation of data from knockout tissues can be complicated by compensatory expression of related proteins, which may show cross-reactivity with the primary antibody being used. In evaluating a novel antibody for immunochemical staining, you should also consider testing the antibody in immunoprecipitation and Western blotting. The detection of minimal background bands in Western blotting and ideally a band at the correct molecular weight gives some reassurance that the antibody has a degree of specificity. Not all antibodies work in all techniques though, so a negative result in a Western blot does not indicate that the antibody will not work in immunochemical applications. Immunoprecipitation relies on the ability to detect the native (detergent-solubilized) form of the protein, which may more closely mimic immunochemical staining, depending on antigen retrieval and fixation conditions, and may provide further indications of antibody specificity.

Antibody specificity and cross-reactivity are descriptive terms, dependent upon the context of the experiment. They are sometimes used in reference to the range of species that an antibody raised against a conserved protein detects, or to refer to proteins other than the target protein that are detected by the antibody. This can be attributed to similar epitopes in distinct proteins, or to the presence of antibodies in a protein A-purified antibody or sera that do not recognize the protein of interest. Contaminating antibodies such as these can be removed by further affinity-purification steps (see section 2.2.8).

2.2 Antibody production

Antibodies for experimental use are usually purified from the sera of an immunized animal; hence the prominent isotype is IgG, with some IgM. Both monoclonal and polyclonal antibodies are routinely employed in immunochemical studies, and the advantages and disadvantages of these antibody types will be discussed. Background on the techniques used for production of polyclonal and monoclonal antibodies will be provided in sufficient detail to allow the researcher to manage a custom antibody production project. Several of the considerations for custom antibody manufacture can also be applied to antibody sourcing from commercial or academic suppliers.

Production of antibodies for immunoassays begins with an initial immunization that stimulates the immune response, leading to a lag phase of several days required for proliferation of the plasma cells, before significant antibody titers are seen. Immunization leads to an early IgM response (see *Fig. 3*), which decreases rapidly. Individual IgM-producing

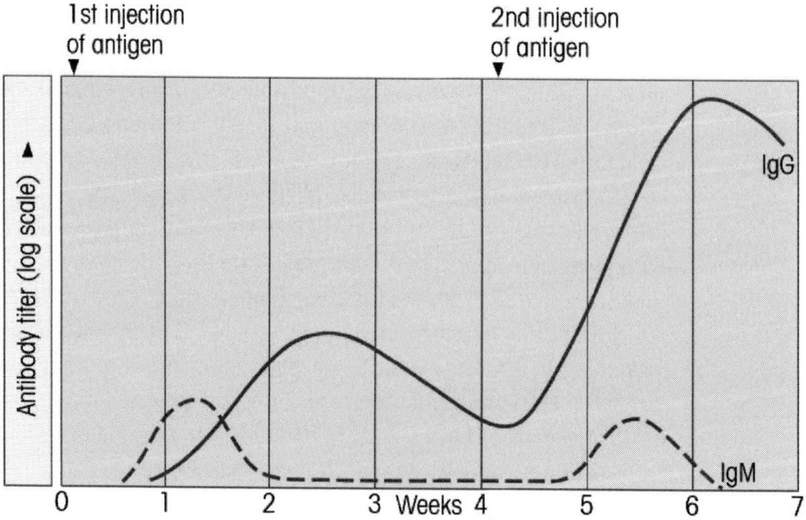

Figure 3. IgM is produced in the early phases of the immune response.
Early responses to antigen include synthesis of IgM. The antigen-recognizing properties of IgM are then converted to form IgG during the T-cell-dependent class switch. Subsequent exposure to the same, or a similar, antigen leads to augmentation of IgG production facilitated by memory B cells. This forms the basis of the extended immunization protocols used in antibody production and is the principle behind vaccination. Reproduced from *Roitt's Essential Immunology* (2001) with permission from Blackwell Publishing.

plasma cells undergo a T-cell-dependent class shift and start to produce IgG with the same specificity. After subsequent injections, the IgG titer increases. This forms the rationale for the extended immunization protocols used for polyclonal antibody production. In addition to the class shift to IgG production, affinity maturation of the antibodies occurs as the immunization protocol is extended. The two main drivers of this are declining immunogen levels, leading to selection of proliferating B cells that produce higher-affinity antibodies, and mutations within the antibody-producing cells, which lead to the formation of variants of the parent antibody, some of which will have a higher affinity for the antigen. This affinity maturation and shift in specificity of antibodies may lead to enrichment of high-affinity antibody, which suits the experimental purpose. The converse can also be true and a population of antibody fit for the purpose can be lost; hence, production bleeds from several stages of the immunization protocol should be screened if you are looking for an antibody with specific characteristics.

2.2.1 Production of polyclonal antibodies

Polyclonal antibodies are available commercially for many protein targets, but where these are not available, or if large quantities are required, the production of custom polyclonal antibodies is cost-effective and straightforward. The success rate of polyclonal antibody production is highly dependent on the application that the antibody needs to work in, immunogen selection and the immunization schedule. Although a workhorse of immunoassays, a limitation of polyclonal antibodies is the potential for significant batch-to-batch variation due to differences in the host immune response to the immunogen. Sufficient antibody of the same batch should be sourced to complete a study and, where possible, an immunogen-affinity-purified antibody should be used to help counter batch-to-batch variation and background staining.

Polyclonal antibodies are commonly produced in mammalian and avian host species. Rabbits are the most widely used host species as they are easy to handle and have a robust immune response to a wide range of immunogens. The inbred New Zealand White rabbit is particularly amenable to antibody production due to its large size, docile nature, and resilience. Sheep, goat, and donkey are also used where large quantities of antibody are required, such as in secondary antibody production, or where antibodies from a different species are required for multiple immunochemical staining studies (see Chapter 5). Guinea pig, mouse, and rat may also be used for polyclonal antibody production, although the yield of antibody is much lower, and antibodies raised in these hosts tend to be used to fulfill specific requirements. Mice tend to have a weaker immune response, and a more restricted range of effective immunogens, than rabbits.

Although secondary antibodies will not be discussed here in detail, the species specificity and lack of cross-reactivity with other mammalian proteins is a key consideration in secondary reagent selection for immunolabelling. Highly cross-absorbed secondary reagents, where the sera is pre-absorbed against immunoglobulins from potentially cross-reacting species, sera proteins, extracts of tissues of interest, or other abundant proteins, are recommended for many immunochemical staining applications. Polyclonal secondary antibodies tend to give a higher degree of amplification than monoclonal secondary antibodies as they recognize a wider range of epitopes, thus increasing avidity effects. It is good practice to test new batches of secondary antibody in parallel with a successful batch, as, like all polyclonal antibodies, they can be affected by batch-to-batch variation.

Chickens are phylogenetically diverse from humans and other mammals, and are useful for raising antibodies against highly conserved mammalian proteins, where antibodies with appropriate specificity may not be in a mammalian host's repertoire due to exclusion of antibodies with potential for self-recognition. They are also a good alternative species for antibody production for multiple immunochemical staining studies. Chicken antibodies are normally harvested from egg yolk, although small amounts of sera can be obtained for pre-purification screening and some antibody can be purified successfully from this. The yolk-derived antibodies are referred to as IgY (yolk) and form the chicken equivalent of IgG in mammalian species. Twelve eggs contain approximately 1 g of IgY, which is equivalent to the IgG content of approximately 100 ml of serum. The amount of antigen-specific antibody varies from 0.5 to 10% of the total IgY. Chicken antibodies from sera are often referred to as IgG, although the fundamental antibody structure is different to that in mammals. The heavy chain of chicken IgY contains four constant domains and one variable domain, but does not bind to Fc receptors or to protein A or G.

If you do not have access to a suitable animal facility and experienced animal technicians, there are several reliable custom polyclonal antibody companies that will recommend production protocols optimized for your particular requirements, such as speed, production of bulk sera, or a specific conjugation strategy. The best custom suppliers will also offer advice on immunogen selection and formulation. It is good practice to ensure that your chosen supplier conforms to legislation regarding animal welfare and ethical use of animals for research work. If you use a custom antibody supplier in a different country, or intend to distribute the resulting antibody to colleagues, it is worth checking on the import/export regulations governing the movements of animal by-products. Many custom suppliers also offer peptide synthesis, carrier protein conjugation, and antibody purification services. The Antibody Resource Page (http://www.antibodyresource.com/) is an excellent starting point when looking for a custom supplier and contains links to other sources of information on antibodies and peptides.

The number of animals used for polyclonal antibody production should, under legal and ethical guidelines, be the minimum required to fulfill the research purpose. However, in order to successfully raise an antibody with the appropriate specificity and affinity, several animals are often required to provide sera of the correct specificity, especially when raising antibodies to post-translational modifications. In order to balance these factors, the use of two to six rabbits is typical, or one to two animals for larger species such as goat. There are several immunization protocols used routinely, and your supplier or animal facility should provide advice on the most appropriate. The important factors in developing a protocol are the host species, the amount and form of the immunogen, the type of adjuvant, the route of injection, immunization and bleed frequency, and the subsequent antibody purification technique. The speed at which the antibody is required often governs the length of the protocol. Longer protocols allow the process of affinity maturation and the collection of larger volumes of sera. However, if a reagent is required urgently, a shortened protocol can be used. Several custom suppliers offer short protocols, but, unless time is very limited, a longer protocol is usually more successful and early bleeds from a longer protocol can be analyzed without compromising the potential to develop a good reagent.

2.2.2 Monoclonal antibody development

Monoclonal antibodies are produced from a hybridoma cell line that secretes antibody of a single idiotype. Monoclonal antibodies are usually IgG, with some IgM available. For ease of purification and handling, IgG is preferable. Monoclonal antibody production is a skilled procedure and mouse monoclonal antibody production facilities are found in many research facilities. Monoclonal antibodies are usually identified by the target and a clone number, for example anti-Myc, clone 9E10. Some commercial suppliers do change the clone numbers of antibodies that they have licensed. However, they should be able to provide additional information if you contact them. Custom monoclonal antibody production is offered by a number of suppliers, but before committing to a custom project, which can be expensive, it is advisable to ensure that the hybridoma development work is done on site at your chosen supplier and not outsourced.

The host species for monoclonal antibody production is usually mouse, although rats may have advantages such as a stronger immune response to a wider range of immunogens and improved hybridoma stability. Recently, rabbit and sheep monoclonal antibodies have come on to the market. The production technology for rabbit and sheep monoclonal antibodies is proprietary and, although the range of targets available is currently limited, the availability of monoclonal antibodies from multiple species should expand in the next few years.

A brief synopsis of a monoclonal antibody production process will follow, but a detailed protocol is beyond the scope of this chapter. Several excellent texts (e.g. 4) have been published on the subject. Mice are immunized with an appropriate immunogen (see section 2.2.4) and an immune response is formed to the multiple epitopes displayed by that immunogen. Antibody titer can be monitored in sera at this point if required, although this is not always representative. Plasma cells from the mouse spleen are then fused in culture with immortal B cells derived from tumors (myelomas) to produce immortalized antibody-secreting hybridomas. After selection of successfully fused cells, tissue culture supernatants from clonal populations of hybridomas are screened for those cells secreting antibody to the antigen of interest, usually using an enzyme-linked immunosorbent assay (ELISA) or flow cytometry. Normally several rounds of screening and subcloning are used to identify single clones with the correct specificity and production capability.

There are two main approaches for monoclonal antibody production from hybridomas. The first, production in ascites, has been the preferred choice historically. Here, hybridoma cells are grown in a mouse and the antibody-rich ascitic fluid is collected from the resulting tumor. This process is under review due to animal welfare guidance and is being phased out by many organizations where possible. Guidelines are in place to restrict production in ascites except where production *in vitro* is not a viable option. Monoclonal antibody production in ascites gives high yields of 1 to >10 mg/ml of specific antibody and this accounts for >90% of the total antibody present in the ascitic fluid. In cases where immunodeficient nude mice are used, the percentage yield is greater. Cell debris is removed from the ascites, which can then be used in a crude form or can be purified using protein A or G. Ascitic fluid contains a number of proteases and is usually stored as frozen aliquots. Sodium azide, or an alternative antimicrobial agent, is usually added, unless the intended use precludes this, such as studies *in vivo*.

The alternative to ascites production is production by cell culture *in vitro*. The major disadvantage of production *in vitro* is that culture supernatants contain a low concentration of antibody (1–100 μg/ml) and, for sufficient yield, purification from large volumes of tissue culture supernatant may be necessary. Protein A purification methods may lead to bovine IgG (from fetal calf serum in the culture supernatant) contaminating the final product, hence the use of low-serum and serum-free media for hybridoma cell culture. Some hybridomas do not adapt well to *in vitro* production conditions.

A major advantage of monoclonal antibodies is the availability of large amounts of a single, well-defined detection reagent. This makes them particularly suitable as cell-type markers and in diagnostics. However, the monospecificity of monoclonal antibodies may limit the range of applications and experimental conditions under which they give good results. Monoclonal antibodies are still prone to nonspecific binding to related epitopes, and good controls and validation are as important for monoclonal antibodies as for polyclonal antibodies. Pooled monoclonal antibodies that have been carefully selected to minimize background cross-reactivity can also be used for immunochemistry and may have some of the advantages of increased signal amplification, as seen with polyclonal antibodies.

2.2.3 Recombinant antibody technology

Recombinant antibody technology allows antibody generation *in vitro* and is at the cutting edge of developments in therapeutic antibodies. It does not as yet provide widespread research reagents, although their availability is increasing. The discovery of techniques to make monoclonal antibodies by Köhler and Milstein in Cambridge in 1974 (5) prompted a wave of development of antibodies as therapeutics. A limitation of mouse monoclonal antibodies as therapeutics is their antigenicity, and early efforts to design therapeutic antibodies focused on humanization and chimerization of mouse monoclonal antibodies to make them more compatible with the human immune system. As a result, both whole IgG and monovalent antibody fragments (Fab) have received regulatory approval. Recombinant antibodies can be derived from libraries constructed from immunized mice, or more commonly now from large libraries of natural and synthetic immunoglobulin genes. Phage display or ribosomal display are used to select fragments with the desired specificity, and further selection and protein engineering are used to improve the specificity, affinity, and pharmacokinetics. The prototypic recombinant antibody fragment is the single-chain Fv (scFv) comprising the V_H and V_L domains joined by a flexible peptide linker. These display similar properties to Fab fragments and, as with Fab fragments, dimerization of scFv increases the avidity of the antibody for antigen and improves its utility. A number of scFv reagents are available for research use, many of which are fused to tags such as Myc or similar and can be detected using anti-tag antibodies. Several biotechnology companies have been founded to explore the potential of smaller antibody fragments, such as single-domain antibodies, diabodies, multispecific recombinant antibodies, and antibodies derived from camelid and cartilaginous fish, which have unique structural features (reviewed in 6).

2.2.4 Immunogens for antibody production

The immune system can respond to a diverse range of molecules and this is reflected in the array of immunogens that have been used successfully for antibody production. The following gives an indication of the classes of immunogenic molecules commonly used to raise antibodies for research purposes. Since the range of potential immunogens is so wide, this is not exhaustive.

Small molecules, or haptens, have been used as immunogens since the 1930s when Karl Landsteiner, one of the pioneers of immunochemistry, used them to investigation antibody recognition at the molecular level and demonstrated the ability of antibodies to discriminate between closely related small molecules (2). He discovered during the course of his studies that small molecules in isolation do not cause an immune response, but when conjugated to an immunogenic carrier protein they lead to the formation of hapten-specific antibodies. This principle is employed today for the generation of large numbers of antibodies to biologically relevant small molecules, such as neurotransmitters and peptides.

Production of a successful antibody is dependent on the selection of an immunogen that contains both B-cell and T-cell epitopes. Carrier proteins are used routinely to provide T-cell epitopes; the target antigen should provide a B-cell epitope. Commonly used carrier proteins are keyhole limpet hemocyanin (KLH), bovine serum albumin (BSA) and diphtheria toxin. Antibodies are raised to the carrier protein in addition to the target immunogen and, as BSA is used widely as a blocking agent for immunological techniques, the presence of anti-BSA antibodies in the sera may be a complication. When handling KLH, a limitation may be the formation of insoluble complexes during conjugation to the immunogen. These are still immunogenic, and when in suspension can still be used effectively for immunization purposes. Where anti-carrier protein antibodies, or other background antibodies, are thought to interfere with the staining, immunogen affinity purification of the antibody is a good solution.

2.2.5 Peptide immunogens for antibody production

Peptide immunogens form the largest class of immunogens used today for production of academic and commercial antibodies to specific protein targets. The advantage of peptides for raising antibodies is the ease and speed of synthesis. Peptide immunogens are usually designed to minimize unwanted cross-reactivity with related proteins and can be used to raise antibodies specific for post-translational modifications such as phosphorylation sites. In a typical peptide of 12–20 residues, there will be a limited number of potential epitopes; hence, polyclonal sera raised against peptides may have characteristics similar to monoclonal antibodies. Peptide immunogens may not recognize the native (or fixative-treated) structure of the protein. However, selection of surface loop regions of the protein during immunogen design increases the possibility of producing an antibody suitable for a wide range of applications. A common and successful strategy is to co-immunize the host with multiple peptide–carrier protein conjugates. Two peptides are frequently selected. Using more than four peptides is not usually recommended, as this is thought to depress the immune response to certain peptides due to 'antigenic competition'. However, due to the complex nature of the immune response, the evidence for this phenomenon is not conclusive. The carrier protein and conjugation methods employed whilst raising the antibodies impact on the properties of the resulting antibody. Hence, optimization of the fixation conditions and antigen retrieval can be crucial to antibody recognition of the target protein in the context of the cell and under the particular experimental conditions. A protocol for peptide immunogen design is given at the end of this chapter. As our understanding of the immune response to peptide immunogens is far from complete, with much based on anecdotal evidence, the design of optimized peptide immunogens is not an exact science.

Peptides and small molecules (<6–10 kDa) require conjugation to a carrier protein (see section 2.2.4). For peptides, a common approach is to include a cysteine residue at the N or C terminus of the peptide to facilitate conjugation to the carrier protein using a heterobifunctional cross-linking reagent. This reacts with lysine residues on the carrier protein and with fully reduced cysteines in the peptide. This method orientates the peptide on the carrier protein and facilitates the construction of columns for subsequent affinity purification of the antibody. The carrier protein is thought to mimic the globular protein and hence for peptides derived from the very C terminus, the cysteine would be added to the N terminus of the peptide and vice versa. For peptides from an internal region of the target protein, the cysteine can be on either end and is normally added to the end of the peptide where the proximal region is predicted to be the most structured.

Where a peptide contains internal cysteine residues, an alternative cross-linking reagent, such as the homobifunctional reagent glutaraldehyde, is used. For peptides lacking internal lysine residues, a terminal lysine can also be added instead of cysteine if preferred. Glutaraldehyde contains two aldehyde groups, each of which reacts mainly with primary amines such as lysine residues and the amino group on the N terminus of peptides. The reaction is less controlled than with heterobifunctional linkers and it is less easy to quantify coupling efficiency. However, the conjugates formed are often highly immunogenic and careful optimization of the pH can be used to bias coupling to the N terminus. Where internal lysine residues are present, interpeptide links are formed, which may under some circumstances mimic the conformation of antigens in fixed samples. However, in some cases this leads to large, multimeric complexes where the peptide is not presented appropriately. Glutaraldehyde bridges often form a portion of an epitope recognized by the immunized animal; therefore, immune serum can be screened in the presence of control glutaraldehyde-coupled peptides, to distinguish sera positive for glutaraldehyde adducts. Glutaraldehyde conjugation is good for small chemical molecules such as neurotransmitters where the molecule–glutaraldehyde forms the epitope. To utilize these antibodies in immunostaining, glutaraldehyde should be included in formaldehyde fixative at about 1% (v/v) to induce formation of the epitope in the fixed samples. Several custom antibody suppliers specialize in producing antibodies tailored towards techniques including specific immunochemical staining protocols and often provide free advice prior to commencement of a project. If neither of the above methods of conjugation is appropriate, peptides can be linked via tyrosine (and to a lesser extent histidine) using the homobifunctional cross-linker bis-diazotized toluidine; or by free carboxyl or amino groups on the C or N terminus; or on lysine, glutamate, or aspartate residues, using carbodiimides.

An alternative to using carrier proteins is to synthesize peptides as multiple antigenic peptides (MAPs). MAPs consist of multiple copies of the peptide immunogen attached to a core structure, usually via lysine residues. MAPs contain four to eight (or more) copies of the peptide and have a weight >10 kDa. A limitation of MAPs is that, during peptide synthesis, the purity of the MAP is difficult to determine. Unless the peptide contains a good T-helper-cell epitope, the success of a MAP may be limited by a lack of affinity maturation and class switch to IgG. However, in certain cases, MAPs can be successful and have the advantage of eliminating the antibody response to the highly immunogenic carrier protein.

2.2.6 Recombinant and purified native protein immunogens

Recombinant full-length proteins, protein fragments, and purified native proteins form very good immunogens, as long as the immunogen is sufficiently different from host proteins to raise a robust immune response. Where protein fragments are used, these can be fused to a tag, such as a hexa-His or glutathione *S*-transferase to facilitate purification and detection. In the case of small tags such as hexa-His or Myc, the tag does not need to be cleaved off before immunization. Immunization with protein gives a true polyclonal response with a wide range of antibodies raised against multiple epitopes. Typically 1–5 mg of purified protein is required for polyclonal production, although protocols can be tailored to use smaller amounts of immunogen. Where proteins cannot be purified in sufficient quantities for standard immunization techniques, the protein sample can be resolved by SDS-PAGE and the band corresponding to the protein of interest excised. The gel slice is then used as an immunogen. This can often give good results and the presence of acrylamide from the gel may boost immunogenicity. A disadvantage with antibodies raised against proteins is the higher likelihood for cross-reactivity with related and unrelated proteins. With any protein immunogen, proving the specificity of the antiserum or affinity-purified immunoglobulin is a challenge.

2.2.7 Other sources of immunogens

Immunogens do not have to be clearly defined entities. Several valuable reagents (the majority of them monoclonal antibodies) have been raised against cells or tissues of interest. Indeed, a worthwhile strategy for raising antibodies to intractable membrane receptor targets is to create a cell line overexpressing the receptor or membrane protein of interest and use these cells, or membrane fractions of these cells, as immunogens. Nucleotides can also be used as immunogens, particularly when complexed with protein.

DNA immunization is a technique used not to raise antibodies to the DNA itself but to use the host's cellular machinery to synthesize protein from the immunized DNA, which is then presented to the host immune system. Several techniques have been used to introduce highly purified plasmid DNA into the host cells, such as injection into host muscle cells and ballistic methods using DNA-coated gold particles, which are introduced into skin cells. Genetic immunization has been used successfully in a number of species including mouse and rabbit. This technique is promising and many laboratories have developed techniques to optimize the efficacy of genetic immunization (*7*). However, at present the success rate is generally lower than classical methods. Genetic immunization is a viable alternative where the target protein is not amenable to other methods of antibody production. Antibodies produced using genetic immunizations are likely to recognize the native form of the protein, as the immunogenic protein is folded within the cells of the host.

2.2.8 Antibody purification: an overview

Although crude sera can be used for many purposes, removal or reduction of sera proteins can be advantageous to improve the specific signal relative to background staining. Preparation of whole IgG fractions from sera using protein A or protein G is a standard technique. Proteins A and G are bacterial proteins that bind immunoglobulins with high specificity. The binding affinity for proteins A and G is dependent on host species and isotype (see *Table 2*). It is also dependent on the salt concentration used. Chicken IgY does not bind protein A or G. The purification of IgY from yolk requires specialized protocols including a delipidation step and the use of protein L. Ammonium sulfate precipitation can also be used to concentrate antibody fractions and, when followed by ion-exchange chromatography, removes the majority of other serum proteins. Detailed discussion of antibody purification is beyond the scope of this book; however, purification methods and suggested protocols can be found in (8).

Polyclonal sera contain unrelated and irrelevant antibodies (90–99% of total IgG) that can increase background staining. For antibodies raised against peptide and protein immunogens, immunogen affinity purification can be used to purify the antibody of interest from the immunoglobulin pool. For immunochemical studies, immunogen-affinity-purified antibodies ensure reproducibility and a clean signal, although the signal specificity is ultimately determined by the quality of the immunogen and the immune response. Immunogen affinity purification is also used to fractionate antibody populations obtained from co-immunization protocols, or where the antibodies are to be labeled with detection molecules or biotin (although protein A- or G-purified antibody can also be used in labeling protocols). Immunogen affinity purification will also remove antibodies to the carrier protein. Multiple peptide affinity purification steps can sometimes be used to remove cross-reactivity with closely related proteins, or, in the case of post-translational modification-specific antibodies, the peptide lacking the modification. Affinity purification may be essential for subsequent immunochemical staining techniques where BSA has been used as a carrier protein.

Some sera may contain high levels of lipid. This should not interfere with successful staining but can be removed using commercial reagents or 1,1,2-trichlorotrifluoroethane. Removal of lipid can also improve subsequent antibody purification but is not essential. Protein A, G, or L purification (with or without lipid removal depending on the circumstances) can be used as an alternative approach if the lipid levels compromise the immunoassay. Hemolysis (red coloration in the plasma due to lysis of red blood cells) is also present in some sera. This is due to suboptimal bleeding techniques or animal-specific factors and is also unlikely to interfere with experimental results. However, one may also like to consider purification under these circumstances.

2.2.9 Antibody stability and storage

Antibodies are stable proteins and with careful handling, the shelf life of an undiluted purified antibody or serum can be more than 10 years. It is, however, good practice to take account of the manufacturer's expiry date for commercial antibodies, although these are often conservative estimates and may be based on accelerated degradation studies. However, expiry dates must be complied with strictly where regulatory guidelines apply, such as in a clinical setting. The stability of antibodies is dependent on the purification method. Exposure to acidic conditions (pH 2–5), used routinely in peptide-affinity purification, can affect antibody stability and potentiate aggregation. This may adversely affect the results obtained in immunochemical staining studies by increasing background staining due to spurious hydrophobic interactions. Conjugation to fluorophores, enzymes, and small molecules can also affect antibody stability, shelf life, and nonspecific background staining. This should be considered when troubleshooting unexpected staining patterns. However, each antibody behaves distinctly and generalization is difficult.

The material of the storage container should minimize passive adsorption to the container. Polycarbonate and polypropylene are often used. Borosilicate glass is also appropriate. Standard glass containers should not be used, as proteins adsorb strongly on to the walls. Where purified antibodies are stored at concentrations of less than 1 mg/ml, the addition of a bulk protein can improve stability. For most applications, the addition of 0.1–1% (w/v) BSA is appropriate. Most antibodies are best stored at -20 to -80°C in a nonfrost-free freezer (many frost free freezers undergo repeated freeze–thaw cycles). Where glycerol is present as a cryoprotectant, or to facilitate handling, antibodies are usually stored at -20°C and at this temperature the antibody solution does not freeze. Antibodies are sensitive to freeze–thaw cycles and should be aliquotted where appropriate. Short-term storage at 2–8°C is also appropriate, although care must be taken to prevent bacterial or fungal growth. Preservatives such as sodium azide (0.002–0.05%, w/v) or thimerisol (0.1%, w/v) are often added. An alternative approach is to filter-sterilize antibody solutions. Sodium azide (NaN_3, sometimes referred to as NaAz) inhibits horseradish peroxidase (HRP) reporter labels and is therefore unsuitable as a preservative for HRP-conjugated antibodies. For situations where azide may interfere with antibody labeling or function, a borate buffer can be used to provide antibacterial protection. Antibodies contaminated by bacterial or fungal growth should not be used, as this will normally interfere with staining or lead to high background levels.

2.2.10 Considerations for antibody sourcing

Antibodies are available from both commercial and academic sources. Where possible, purchase antibodies that have been tested in the specific application you intend to use them for, such as paraffin-embedded or frozen sections, and where recommendations for antigen retrieval and a suggested dilution are given. For established targets, an appropriate source can often be identified from the literature. For novel targets or applications, it is important to purchase from a reputable company with good customer service, as antibodies are notoriously fickle reagents and polyclonal antibodies in particular can be susceptible to batch-to-batch variation. Where an antibody fails to perform as stated in the company's literature, you should be provided with constructive feedback from the company and recompense if the quality of the antibody is in doubt.

Antibodies from academic sources that have been referenced in peer-reviewed articles should be available to others working in the field. In some countries, a Materials Transfer Agreement or equivalent may need to be provided, which may limit the scope of use for that particular antibody. Researchers are expected to supply antibodies in reasonable amounts sufficient to allow the original data to be reproduced and extended. Many academics with popular antibodies license these to commercial suppliers or, in the case of monoclonal antibodies, deposit them in a hybridoma bank or provide the clone to facilitate the distribution of these antibodies where there may be an overwhelming number of requests for material.

Monoclonal antibodies are more easily sourced, as the antibody produced by that clone should be consistent across academic and commercial sources. Again, source directly from the originator where possible, or, if the clone has been licensed, buy from a reliable supplier. Where a large amount of monoclonal antibody is required for a study, it is worth checking whether the clone is freely available, either from the originator or from the American Type Culture Collection, the European Collection of Animal Cell Culture or another hybridoma bank. Outsourced monoclonal antibody production can be a cost-effective way of producing large quantities of antibody if in-house facilities are not available. It is good practice to ensure that hybridoma lines are free of mycoplasmas.

2.3 Recommended protocols

Protocols for antibody production, peptide conjugation, and affinity purification can be found in several laboratory manuals (e.g. 4, 8–10) and are beyond the scope of this chapter. Protocols for peptide immunogen design are less well documented. A contributor to this is the lack of understanding of determinants for good peptide immunogens (8). The following protocol has given good results in our hands. However, there are no definitive solutions, so the protocol can be adapted to take into account additional factors.

2.3.1 Peptide immunogen design

Peptide immunogens are an accessible way for researchers to design antibodies specific to their target protein and to purify the resulting antibodies, hence limiting nonspecific interactions. Care with peptide selection can lead to the generation of high-affinity, specific antibodies. The following protocol utilizes resources in the public domain, and widely available commercial software, to present an easy approach to peptide immunogen design. Aspects of the following protocol may not be relevant to researchers working on less-well-studied species due to a dearth of database information. However, the principles are generally applicable. Several custom polyclonal antibody companies offer consultation at the immunogen design stage. It should also be noted that, for the same reasons, the following protocol is unable to follow the standard format of the other protocols of this book.

Resources required

- Computer with internet connectivity
- Software for hydrophilicity, antigenicity, and secondary protein structure prediction, or access to internet-based prediction software

The principles of peptide selection (applicable to the design of all peptides)

- Peptide immunogens should be selected that lie in areas that are likely to be surface exposed; these are usually hydrophilic areas.
- Peptides should correspond to loops in the parent protein to best mimic the presentation of the sequence in the native protein.
- Avoid areas that give unwanted cross-reactivity with related proteins and sequences that contain potential post-translational modifications.

- As a general rule, the C and N termini (but ensure that the extreme N-terminal region is not cleaved off) should be considered first as these tend to be surface exposed and antibodies raised against the termini have a good success rate in immunochemical staining and in Western blotting.

1. Identify a reliable protein sequence and note features of the sequence.
   - It is good practice to check your protein sequence against a database. The SwissProt database (http://www.expasy.ch/sprot/) is a good starting point for many targets. Many database entries will contain information on isoforms, sequence conflicts, protein topology, and post-translational modifications.
   - Note areas to avoid from as many database entries as are available for your protein of interest. Other sites for gathering information on your protein are those that collect information from several web resources such as EMBL Harvester (http://harvester.embl.de/). Note the locations of single amino acid substitutions, glycosylation sites, phosphorylation sites, conflicts, transmembrane regions, signal peptides, cleavage sites, and other post-translational modification sites. For less-well-characterized proteins, potential modifications can be predicted using tools available in the public domain such as those on the ExPASy Tools website (http://us.expasy.org/tools/).
   - For projects with multiple isoforms, or those within a family of conserved proteins, a primary sequence alignment is helpful in identifying regions that are distinct from closely related proteins or regions conserved between isoforms and helps to highlight regions to consider for potential immunogens. Clustal w is a reliable and easy-to-use public domain alignment program (http://www.ebi.ac.uk/clustalw/).

2. Identify and note down candidate immunogenic peptides.
   - Identify regions of the protein that are likely to be surface-exposed loops. If the tertiary structure is not available, a number of tools that predict protein secondary or tertiary structures can be used. These include commercial programs and web resources. Several algorithms are used for secondary structure prediction, the most popular being Chou–Fasman. Secondary structure prediction can be inaccurate and should be treated with caution.
   - Also consider the C and N termini as these are frequently surface exposed.
   - Hydrophilicity is frequently used as a measure of potential surface exposure and hydrophilic peptides are often good immunogens and are easier to synthesize and handle. Note the hydrophilicity of your candidate peptides and use this in conjunction with secondary

structure predictions to identify potential surface-exposed loops if no tertiary structure is available. The hydrophilicity profile of the protein can be predicted using the approach of Kyte and Doolittle (11), which predicts regional hydropathy of proteins from the amino acid sequences, or the method of Hopp and Woods (12), which is used to predict protein antigenic determinants by searching protein sequences for the area of greatest local hydrophilicity.

- The antigenic index for protein sequences can be also calculated using algorithms such as those of Jameson and Wolf (13), which predict potential antigenic determinants using an index of antigenicity based upon hydrophilicity (9), surface probability (14), flexibility (15), and the secondary structure predictions of Chou and Fasman (16) and Garnier and Robson (17).
- Narrow down your candidate immunogens by including only hydrophilic regions of greater than ten residues with a high antigenic index and that are not predicted to be in α-helices or β-sheets.

3. Peptide selection: apply the following criteria to your remaining candidate peptides.
   - Exclude those with repetitive sequence (three or more identical residues). These may lead to poor peptide quality or nonspecific antibodies. In particular, three or more consecutive D (aspartic acid) residues may form aspartimides and so should be avoided where possible. Also avoid poly(Q) (glutamine) and peptides containing more than three Q residues across the length of the sequence, as hydrogen bonding can occur between peptides and lead to poor peptide synthesis and low solubility.
   - Within reason, avoid large numbers of M (methionine), R (arginine), W (tryptophan), and C (cysteine) residues, as these may cause problems in peptide synthesis and may form side reactions or be oxidized.
   - Avoid blocks of hydrophobic residues (greater than four residues). This is important for immunogenicity, peptide synthesis, and solubility.
   - Consider avoiding two or more adjacent S (serine), T (threonine), A (alanine), V (valine), or Y (tyrosine) residues where possible, as these may cause problems in peptide synthesis. However, most peptide suppliers have strategies to deal with these, so do not exclude otherwise good potential immunogens for this reason.
   - Y and P (proline) are good residues to include in an immunogen, as the bulky ring structures gives structure to the peptide.

- Look for a balance of charge, with a net charge of <3. Where possible, avoid long blocks of charge.

4. Check any remaining candidate peptide for cross-reactivity with related and unrelated proteins.
   - Using an algorithm and settings suitable for short peptide sequences (such as the NCBI short, nearly exact BLAST search). If a sequence is likely to cross-react with related or highly abundant proteins, discount it. Judging potential cross-reactivity is subjective as the nature of the epitope will be unknown. Blocks of amino acids of five residues or more, especially near the ends of the peptide, should be avoided if these are present in other proteins from the species of interest. You may find that restricting the search to the species of interest or mammalian proteins alone helps with interpretation of the BLAST results. This can be done before the 'format' button is clicked on the NCBI site. Note those immunogens with good species cross-reactivity if this is relevant to your research.

5. Refine your chosen peptide(s).
   - The optimum peptide length is around 15 residues. A length of 12–15 residues is good for N- or C-terminal peptides. For internal peptides, a slightly greater length can be used (up to 20 residues is typical) to allow for multiple epitopes. The final length will be determined by the peptide sequence and cross-reactivity with other proteins.
   - Additional residues for conjugation or modification purposes may need to be added at this point.

6. Optional: peptide immunogen design for recognition of post-translational modification sites.
   - Immunogens to recognize post-translation modification sites tend to have the modification in the center and the maximum length of the immunogen should be about 15 residues. More than ten residues is recommended; however, post-translational modification-specific antibodies have been raised against both shorter and longer sequences. As a general guide, place the modified residue in the center of the peptide and extend the sequence by five residues in either direction.
   - In some circumstances, a GG (double glycine) spacer can be added between the residue added for conjugation purposes and the native sequence to compensate for very short immunogens, which may be needed to give the required specificity.
   - Use the results of BLAST searches to eliminate cross-reactive sequences and shorten or extend the relevant ends of the sequence as required. Although extending the sequence beyond seven to nine

residues either side of the modification may promote the dominance of antibodies against the unmodified form of the immunogen, antibodies towards unmodified forms can be removed by depletion during affinity purification.

- Synthesize the peptide without the modification for use in purification protocols and as a control for specificity in ELISA and other immunoassays.

3. REFERENCES

1. **Roitt IM & Delves PJ** (2001) *Roitt's Essential Immunology*, 10th edn. Blackwell Science, Oxford, UK.
2. **Landsteiner K** (1945) *The Specificity of Serological Reactions*. Harvard University Press, Cambridge, MA, USA.
3. **Braden BC & Poljak RJ** (1995) *FASEB J.* **9**, 9–16.
4. **Shepherd PS & Dean C** (eds) *Monoclonal Antibodies: a Practical Approach*. Oxford University Press, Oxford, UK.
5. **Köhler G & Milstein C** (1975) *Nature*, **256**, 495–497.
6. **Holliger P & Hudson PJ** (2005) *Nat. Biotechnol.* **23**, 1126–1136.
7. **Chambers RS & Johnston SA** (2003) *Nat. Biotechnol.* **21**, 1088–1092.
8. **Harlow E & Lane D** (eds) (1988) In *Antibodies: a Laboratory Manual*, pp. 616–619. Cold Spring Harbor Laboratory Press, NY, USA.
9. **Goding JW** (1996) *Monoclonal Antibodies: Principles and Practice: Production and Application of Monoclonal Antibodies in Cell Biology, Biochemistry and Immunology*, 3rd edn. Academic Press, London, UK & New York, USA.
10. **Liddell JE & Cryer A** (1991) *A Practical Guide to Monoclonal Antibodies*. John Wiley & Sons, Chichester, UK.
11. **Kyte J & Doolittle RF** (1982) *J. Mol. Biol.* **157**, 105–132.
12. **Hopp TP & Woods KR** (1981) *Proc. Natl. Acad. Sci. U. S. A.* **78**, 3824–3828.
13. **Jameson BA & Wolf H** (1988) *Comput. Appl. Biosci.* **4**, 181–186.
14. **Emini EA, Hughes JV, Perlow DS & Boger J** (1985) *J. Virol.* **55**, 836–839.
15. **Karplus PA & Schulz GE** (1985) *Naturwissenschaften* **72**, 212–213.
16. **Chou PY & Fasman GD** (1978) *Adv. Enzymol. Relat. Areas Mol. Biol.* **47**, 45–148.
17. **Garnier J, D. Osguthorpe J & Robson B** (1978) *J. Mol. Biol.* **120**, 97–120.

CHAPTER 3
The selection of reporter labels

Sarah Mardle

1. INTRODUCTION

Immunochemical staining methods have evolved rapidly over the past 100 years and there is now a wide range of choice of detection systems available.

By using a reporter label, researchers are able to visualize an antibody bound to the protein of interest in a tissue section or cell. The label can be enzymatic or fluorescent in nature, bound directly to the antibody raised against the target protein ('primary antibody'), bound to a 'secondary antibody', which recognizes the primary antibody, or bound to a tertiary antibody or suitable amplification system (see Chapter 4, section 2.10.2).

The first immunochemical stains used secondary antibodies conjugated to powerful enzymes. When exposed to the appropriate substrate (chromogen), they were capable of producing striking and colorful precipitates visible under a light microscope that, in some preparations, could even be seen by the naked eye. This was the golden era of the use of horseradish peroxidase (HRP) and alkaline phosphatase (AP). Mastering a good immunochemical stain with an enzyme-conjugated secondary antibody is still very much a respected and widely used technique in anatomical studies today.

The recent advances in fluorochrome-conjugated secondary antibodies has enabled a new generation of brightly colored images and the endless possibilities of double and triple stains due to the fact that a fluorochrome absorbs and emits light in a specific spectrum seen under a specific filter. Several fluorochromes, therefore, can be used at the same time and observed in individual filters, enabling scientists to study further into the biological processes detectable by immunochemical staining (see Chapter 5, section 1.2.2).

In the most recent years, scientists have been able to produce high-quality immunochemical stains due to the development of:

Immunohistochemistry: *Methods Express* (S. Renshaw, ed.)

- Specific antibodies giving even lower background (see Chapter 2).
- Highly powerful fluorochromes (the Alexa generation).
- New amplification kits to detect minute amounts of protein (previously not detectable by direct immunochemical staining) by using compounds that have affinities to complex with each other (streptavidin, biotin, extra-avidin) and can be conjugated to both enzymes (for example, streptavidin–HRP) and fluorochromes (for example, streptavidin–fluorescein isothiocyanate (FITC)) (see Chapter 4, section 2.10.2).

This chapter will cover the basic principles of reporter label detection systems and will help the reader choose the correct one for their experiments.

2. METHODS AND APPROACHES

2.1 Enzymatic/chromogenic detection

Secondary or tertiary antibodies, or avidin/avidin–biotin complexes, conjugated to enzymes via covalent or noncovalent chemical methods are incubated with tissue sections or cells following application of the primary antibody to bind specifically to the target antigen. Please note that primary antibodies can be directly conjugated to enzymatic labels, although signal amplification may be a problem if the antigen in question is present in low concentrations (see Chapter 4, section 2.10.2). Following washing to remove excess reagents, the tissue or cells (and therefore the conjugated enzyme labels) are exposed to an excess of chromogen (see Chapter 4, section 2.10.2). The enzyme label will convert the chromogen into a stable precipitate of a specific color, localized at the site of the enzyme label (and therefore antibody binding) and visible by light microscopy:

Enzyme + substrate (chromogen) $\rightarrow$ enzyme + product (precipitate)

For example, the addition of the chromogen 3,3′-diaminobenzidine tetrahydrochloride (DAB)–nickel to a tissue pre-incubated with an HRP-conjugated antibody will produce a strong black/brown-colored precipitate (product) where the secondary antibody (and therefore indirectly the primary antibody and the protein of interest) is present (see *Fig. 1*, also available in the color section).

There are several enzymes that can be used as labels in immunochemistry, most having several choices of chromogen. It is the chromogen that determines the color of the precipitate. The choice of color is largely dependent on personal preference, although all

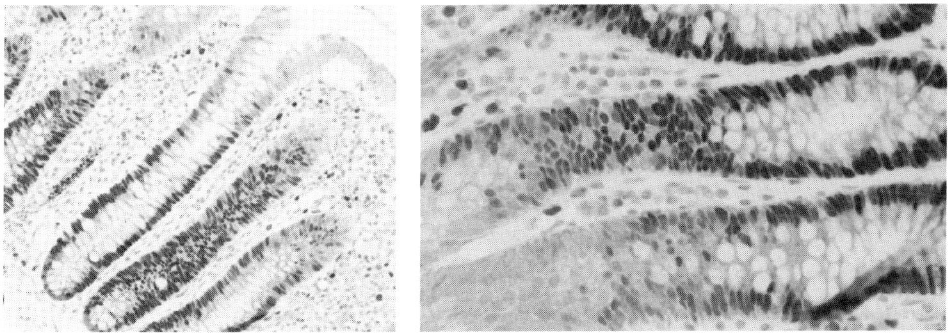

(*a*) (*b*)

Figure 1. Immunochemical staining of human colon by a primary antibody raised against MCM 2 (see page xiii for color version).
The secondary antibody was conjugated to HRP and the chromogen used was DAB (see *Table 1*). MCM 2 is a marker of cellular proliferation; hence, positive staining is predominantly demonstrated in the colonic crypts. Magnification: (*a*) ×20; (*b*) ×60.

chromogens have certain properties that make them more or less favorable depending on the given situation (see below). The various enzymatic label/chromogen combinations are summarized in *Table 1*.

Table 1. Enzymatic labels and their corresponding chromogens

Enzyme	Chromogen	Color of precipitate
Horseradish peroxidase (HRP) (*popular*)	AEC (3-amino-9-ethylcarbazole) (*popular*)	Red
	DAB (3,3′-diaminobenzidine tetrahydrochloride) (*popular*)	Brown
	DAB with nickel enhancement (*popular*)	Black
	CN (4-chloro-1-naphthol) (*unpopular*)	Blue
Alkaline phosphatase (AP) (*popular*)	Naphthol AS-MX phosphate *plus*	
	• Fast blue (*popular*) *or*	Blue
	• Fast red (*popular*)	Red
	Naphthol AS-BI phosphate *plus*	
	• New fuchsin (*popular*)	Red
	BCIP (5-bromo-4-chloro-3-indoxyl phosphate, *p*-toluidine salt) *plus*	
	• INT (*p*-iodonitrotetrazolium chloride) (*unpopular*) *or*	Brown
	• NBT (nitroblue tetrazolium chloride) (*unpopular*) *or*	Deep purple
	• TNBT (tetranitro blue tetrazolium chloride) (*unpopular*)	Deep purple
Glucose oxidase (*unpopular*)	MPMS (methoxyphenazine methosulfate) *plus*	Deep purple
	• NBT (nitroblue tetrazolium chloride)	Deep purple
β-Galactosidase (*unpopular*)	X-gal (5-bromo-4-chloro-3-indolyl-β-galactopyranoside)	Blue

2.1.1 Peroxidase chromogens

Peroxidase chromogens are widely used for both immunoblotting and immunochemical staining procedures and react with the enzyme HRP isolated from the root of the horseradish plant.

In the presence of a suitable electron donor, such as the chromogen DAB, HRP initially forms a complex with hydrogen peroxide, catalyzing its breakdown into water and oxygen. This reaction is driven by the electron donated from the oxidation of the chromogen, the oxidized chromogen forming a colored precipitate at the site of the reaction. An excess of chromogen will reversibly inhibit the enzyme, as well as chemicals such as methanol, azide, and cyanide. Care should be taken to prevent such inhibition, for example when using azide as a preservative in antibody storage buffers (1).

3-Amino-9-ethylcarbazole (AEC)

AEC produces a red precipitate that is soluble in alcohol (use an aqueous mounting medium. See Chapter 4, section 2.7 for further information on mounting media). The intensity of AEC staining is reduced when left in the light over long periods. Storage of AEC-developed slides in the dark is therefore encouraged.

Diaminobenzidine (DAB)

The most commonly used HRP chromogen is DAB (1). DAB produces a brown- or black-colored polymerized precipitate (depending on whether or not nickel enhancement is used) that is insoluble in alcohol (see *Fig. 1*, also available in the color section). Slides should be mounted in an appropriate organic mounting medium following a stringent dehydration procedure to remove all traces of water from the tissue. DAB is a carcinogenic chemical so care must be taken at all times to handle it properly with gloves and on protected benches. DAB is deactivated by incubating the solution and all contaminated laboratory equipment in 10% (v/v) bleach in water for 24 h.

4-Chloro-1-naphthol (CN)

CN produces a blue-colored precipitate that is soluble in alcohol (use an aqueous mounting medium). The disadvantage of this chromogen is that it has a tendency to diffuse away from the site of initial precipitation during storage.

2.1.2 Phosphatase chromogens

Phosphatase chromogens are also widely used for both immunoblotting and immunochemical staining procedures and react with the enzyme AP isolated from calf intestine (1).

AP hydrolyses and therefore removes naphthol phosphate groups from organic esters (e.g. BCIP), creating phenolic compounds. These compounds then couple to diazonium salts (e.g. INT, NBT, or TNBT), reducing them to form a colored precipitate (formazan) at the site of the reaction. Important enzyme co-factors for this reaction are Mn^{2+}, Ca^{2+}, and Mg^{2+}.

Naphthol AS-MX phosphate/Fast blue or Fast red

This reaction produces either a bright blue or red precipitate that is soluble in alcohol (use an aqueous mounting medium).

Naphthol AS-B phosphate/New fuchsin

This reaction produces a bright red precipitate, more intense than that of Fast red. The precipitate is also alcohol insoluble, permitting the use of organic mounting media.

BCIP/INT

This reaction produces a brown precipitate that is soluble in alcohol (use an aqueous mounting medium).

BCIP/NBT or TNBT

Both of these reactions produce a purple precipitate that is alcohol insoluble, permitting the use of organic mounting media. BCIP/TNBT produces a deeper-colored purple precipitate than that of BCIP/NBT.

2.1.3 Glucose oxidase

Glucose oxidase is very rarely used as a reporter label, but is included here for completeness. When coupled to NBT, the chemical MPMS is used as an intermediate electron carrier, reducing NBT to produce a deep purple precipitate. Use an organic mounting medium.

2.1.4 β-Galactosidase

This enzyme is seldom used but is included for completeness. Its reaction with X-gal produces a blue precipitate.

2.2 Fluorescence detection

Secondary or tertiary antibodies, or avidin/avidin–biotin complexes, conjugated to fluorochromes (fluorescent molecules/labels) via covalent or non-covalent chemical methods, are incubated with tissue sections or cells following application of the primary antibody to bind specifically to the target antigen (see *Fig. 2*, also available in the color section). Please note that primary antibodies can be directly conjugated to fluorescent labels, although lack of signal amplification may be a problem if the antigen in question is present in low concentrations (see Chapter 4, section 2.10.2). Following immunochemical staining, the tissue or cells (and therefore the conjugated fluorochromes) are viewed using a fluorescence microscope set up according to the excitation and emission characteristics of the label being used (see *Table 2*) (2).

Fluorescence is the property of atoms or molecules to absorb light of a specific wavelength and to re-emit light at longer wavelengths (see *Fig. 3*). The emission of light at longer wavelengths is due to the loss of energy by the molecule during interactions with its environment prior to the emission of fluorescence, a phenomenon known as internal conversion (the loss of energy in the absence of the emission of light). To understand fluorescence, the electronic states of fluorochromes must be considered.

Electrons exist in a ground state (S_0) or in excited states of higher energy (S_1, S_2). At each of these electronic levels, fluorochromes can exist in a number of vibrational levels (0, 1, 2). At room temperature, thermal

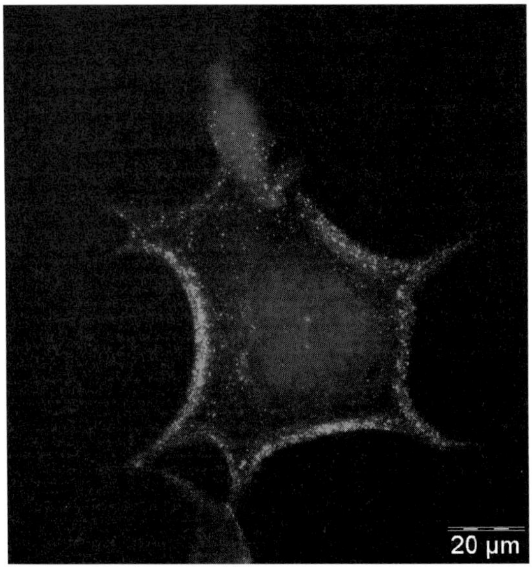

Figure 2. Immunofluorescent staining of human HeLa cells using a monoclonal antibody to pan-cadherin (see page xiii for color version).
The secondary antibody (green) was conjugated to Alexa Fluor 488. DAPI was used to stain the cell nuclei (blue). Alexa Fluor 594 conjugated to phalloidin was used to label F-actin (red).

Table 2. Properties of common fluorochromes

Dye	Absorbance wavelength (nm)	Emission wavelength (nm)	Visible color
Hydroxycoumarin	325	386	Blue
Methoxycoumarin	360	410	Blue
Alexa Fluor 350	345	442	Blue
Aminocoumarin	350	445	Blue
Cy2	490	510	Dark green
FAM	495	516	Dark green
Alexa Fluor 488	494	517	Light green
Fluorescein FITC	495	518	Light green
Alexa Fluor 430	430	545	Light green
Alexa Fluor 532	530	555	Light green
HEX	535	556	Light green
Cy3	550	570	Yellow
TRITC	547	572	Yellow
Alexa Fluor 546	556	573	Yellow
Alexa Fluor 555	556	573	Yellow
R-phycoerythrin (PE)	480, 565	578	Yellow
Rhodamine Red-X	560	580	Orange
Tamara	565	580	Red
Cy3.5	581	596	Red
Rox	575	602	Red
Alexa Fluor 568	578	603	Red
Red 613	480, 565	613	Red
Texas Red	595	615	Red
Alexa Fluor 594	590	617	Red
Alexa Fluor 633	621	639	Red
Allophycocyanin	650	660	Red
Alexa Fluor 633	650	668	Red
Cy5	650	670	Red
Alexa Fluor 660	663	690	Red
Cy5.5	675	694	Red
TruRed	490, 675	695	Red
Alexa Fluor 680	679	702	Red
Cy7	743	770	Red

energy is insufficient to populate the excited electronic states (S_1, S_2) or the higher vibrational states of the ground state. Therefore, absorption generally occurs from molecules in the lowest vibrational energy state and only in the presence of light. Following absorption, a fluorochrome is usually excited to a higher vibrational level of either S_1 or S_2. Subsequently, fluorochromes relax to the lowest vibrational level of S_1, completing the process of internal conversion. Relaxation from this state to the ground state with the emission of a photon is what is known as fluorescence. Each fluorochrome can repeat the excitation/emission cycle

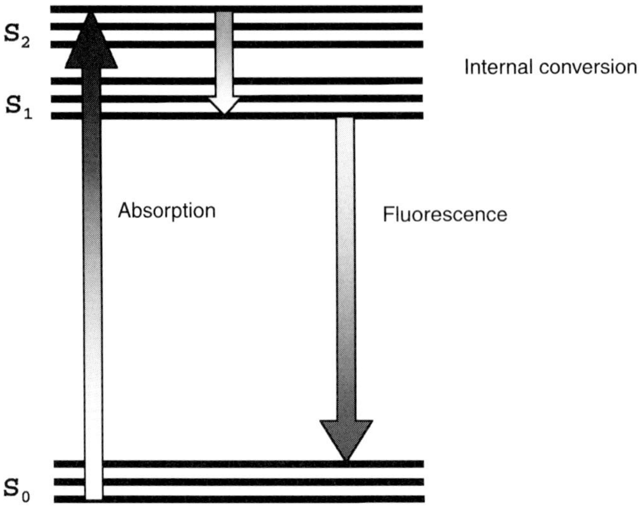

Figure 3. Electronic states and fluorescence.
Electrons can exist in a ground state (S_0) or in excited states of higher energy (S_1, S_2). Absorption typically occurs from molecules in the lowest vibrational energy state. A fluorochrome is usually excited to a higher vibrational level of either S_1 or S_2. Subsequently, fluorochromes relax to the lowest vibrational level of S_1, completing the process of internal conversion. Relaxation from S_1 to the ground state with the emission of a photon is physically what is referred to as fluorescence. Adapted from (3).

numerous times before excitation bleaches the fluorescent signal (photobleaching). For example, FITC can repeat the excitation/emission process approximately 30 000 times before photobleaching occurs. New-generation fluorochromes such as quantum dots (QDots) (4) and the Alexa Fluor range do not photobleach as rapidly as FITC and are brighter and have a broader excitation spectrum, therefore requiring a single source of excitation. The properties of common fluorochromes are given in *Table 2*.

2.3 Blocking unwanted background signals

Undesirable nonspecific 'background' staining can often be seen following staining with both enzymatic and fluorescent techniques. Nonspecific staining patterns do not represent the target antigen and can be due to various elements in tissue sections or cytological preparations. Please see Chapter 9, section 2.8, for a detailed discussion.

2.4 Chromogen development

Chromogen development time can be very varied depending on the amount of target protein. A tissue containing a large amount of target protein or stained using an amplification system is likely to give a strong signal after 1–5 min, due to the abundance of label-bound reagents. A tissue containing a small amount of target protein and stained with no amplification can be expected to require incubation for up to 20–30 min. It is therefore recommended to have a light microscope available during the development stage so that sections can be observed regularly until the desired degree of development is achieved.

2.5 Mounting media and slide storage

Following immunochemical staining, mounting medium is used to adhere the coverslip (a thin piece of glass) to the tissue section or cytological preparation. This is predominantly in order to protect the specimen and the immunochemical staining from physical damage, but also importantly to help to improve the clarity and contrast of the image during microscopy. Several types of mounting medium are available, each with different properties and therefore selectively favoring each of the various chromogens and fluorochromes. Chapter 4, section 2.7 considers mounting media in detail, whilst section 2.1 of this chapter indicates whether a particular chromogen requires an organic or an aqueous mounting medium.

Certain chromogen precipitates fade in sunlight, as do fluorochromes. Similarly, some mounting media can discolor with age. These and other issues regarding slide storage are discussed in detail in Chapter 4, section 2.8.

2.6 Storage of enzyme- and fluorochrome-conjugated reagents

Both enzyme- and fluorochrome-conjugated reagents are best stored in the dark at 4°C. A domestic refrigerator is ideal for this. Darkness protects fluorochromes from becoming photobleached, and a temperature of 4°C prevents ice crystals from damaging labels of either type. If 4°C is not possible, then adding glycerol to a final concentration of 50% and storing at –20°C is an option. The glycerol acts as an antifreeze, hence protecting against ice-crystal damage. However, freezing temperatures are best avoided if possible.

2.7 Enzymatic or fluorescent?

When faced with such a varied choice of labels, it is difficult to know which is best for your purpose. The following is a summary of the advantages and disadvantages of enzymatic and fluorescent labels.

2.7.1 Enzymatic detection

Advantages:

- Enzyme-conjugated reagents do not have to be stored in the dark.
- Incubations of any enzyme-conjugated reagents do not have to be performed in the dark.
- Signal amplification is easy to achieve with the wide range of kits now available.
- A specialist (fluorescence) microscope is not required to view the results.
- Stained tissue can be stored for a long time and will not lose color/intensity significantly.
- Organic mounting media can be used with some chromogens, giving a crisper and sharper image under the microscope.

Disadvantages:

- Multiple immunochemical staining is harder to visualize, as different-colored precipitates may not be as distinctive as fluorochromes, for example double staining with DAB (black or brown) and BCIPI/TNBT (dark purple).
- Double staining for proteins located in the same subcellular compartment is not advisable since the first chromogen precipitate 'fills' the compartment and does not enable the second chromogen precipitate to fill the same compartment adequately. Similarly, co-localization may be difficult to visualize.
- There may be unwanted background staining due to endogenous peroxidases, phosphatases, and endogenous biotin that have not been correctly blocked.
- Health hazards: some chromogens are carcinogenic, such as DAB.
- Chromogens must be made up correctly in order to achieve optimal precipitate development and must not contain any enzymatic inhibitors, such as sodium azide in the case of HRP.

2.7.2 Fluorescence detection

Advantages:

- Multiple staining techniques give excellent color contrast between fluorochromes (see *Fig. 4*, also available in the color section).
- Multiple staining for two proteins located in the same subcellular compartment is not a problem since fluorochromes will absorb/emit light within a specific range and so can be separated easily (see Chapter 5). Overlapping of two fluorochromes can be separated by chromatic filters/computer imaging techniques and reveal proteins in the same cell or in a subcellular compartment.

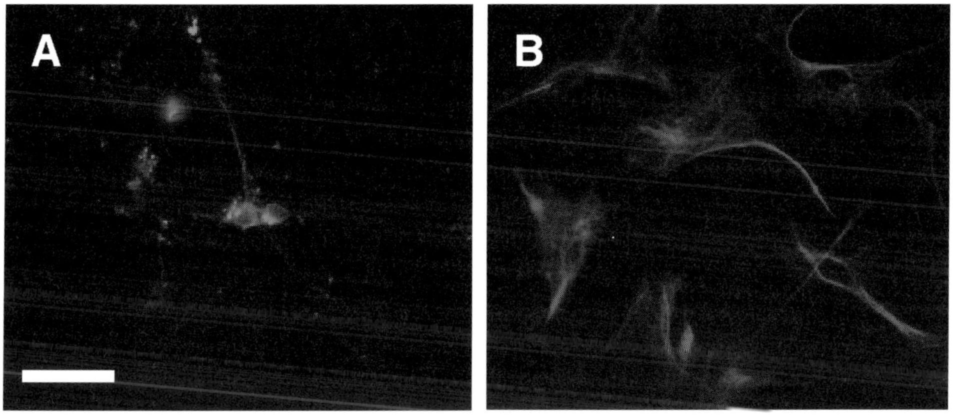

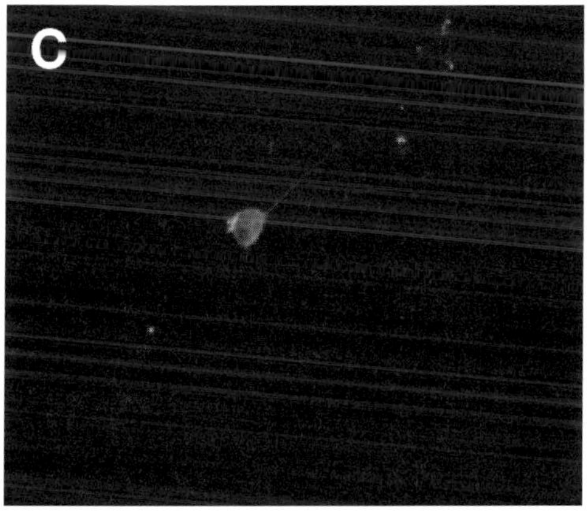

Figure 4. Immunofluorescence enables triple color images to be taken from the same sample (see page xiv for color version).
Here, a mixed population of cells can be distinguished comprising cultured neurons (stained with a primary antibody raised against a neuron-specific protein, detected with a secondary antibody conjugated to an FITC label, visualized at 518 nm) (*A*), glia (stained with a primary antibody raised against a glia-specific protein, detected with a secondary antibody conjugated to a Cy3 label, visualized at 570 nm) (*B*), and microglia (stained with a primary antibody raised against a microglia-specific protein, detected with secondary antibody conjugated to an Alexa Fluor 350 label, visualized at 442 nm) (*C*)

- Staining can be very rapid as fluorochromes are often directly linked to a secondary antibody.
- Confocal microscopy is possible, offering more information (see Chapter 6).

Disadvantages:

- Fluorescence is photobleached very rapidly, a term called 'quenching'. Staining experiments must therefore be done in the dark and quantification/image acquisition must be done in a matter of days after the staining procedure.
- Storage conditions are complex: storage temperature and choice of mounting media is important to retain high-quality staining.
- Amplification is harder to achieve to the same degree as with enzymatic labels, but is possible with the wide range of kits now available.

3. REFERENCES

★★ 1. **Boenisch T (ed.)** (2001) *Immunohistochemical Staining Method*, 3rd edn. Dako Corporation, Carpinteria, CA, USA. – *A detailed guide containing essential information on how best to perform IHC with chromogenic detection.*

2. **Polak JM & van Noorden S** (1997) *Introduction to Immunocytochemistry*, 2nd edn. RMS Microscopy Handbook 37. Bios Scientific Publishers, Oxford, UK.

3. **Herman B** (1998) *Fluorescence microscopy*, 2nd edn. Bios Scientific Publishers, Oxford, UK, in Association with the Royal Microscopical Society.

4. **Michalet X, Pinaud FF, Bentolila LA, *et al.*** (2005) Quantum dots for live cells, in vivo imaging, and diagnostics. *Science*, **307**, 538–544.

CHAPTER 4

Immunochemical staining techniques

Simon Renshaw

1. INTRODUCTION

Immunochemistry is the identification of a certain antigen in a histological tissue section or cytological preparation via an antibody specific to the antigen. The localization of the primary antibody (and therefore the target antigen) is then visualized microscopically via an appropriate enzymatic or fluorescent detection system. There are numerous techniques available, depending on the complexity and on the degree of sensitivity required.

To set the scene, there are basically five types of specimen commonly used in immunochemical staining experiments:

- Paraffin-embedded tissue sections.
- Frozen tissue sections.
- Free-floating tissue sections.
- Cytological specimens as traditional smears.
- Cytological specimens as monolayer preparations (including cytospin preparations).

Although this book is not primarily concerned with the technicalities of physically producing specimens, it is worth outlining the key stages of producing each for completeness (1). The various issues below concerning the preparation, fixation, and storage of specimens in relation to immunochemical staining will be discussed in greater detail later in this chapter.

1.1 Paraffin-embedded sections

Paraffin-embedded sections are probably the most common type of specimen to be immunochemically stained. Tissue blocks are obtained from

Immunohistochemistry: *Methods Express* (S. Renshaw, ed.)

a suitable gross specimen, fixed to preserve morphology, and then processed to paraffin wax in order to give the tissue support during microtomy. Sections are cut at around 4 μm thickness and floated out in a water bath before being picked up on to a glass microscope slide. The sections are then dried at around 60°C in order to increase their adherence to the slide and to help to iron out any creases obtained during microtomy or floating. Before immunochemical staining can occur, sections must be dewaxed and any necessary pre-treatments performed.

1.2 Frozen tissue sections

Frozen tissue sections tend to be utilized when a rapid diagnosis is required or when certain antigens need to be visualized that are sensitive to aldehyde fixation and tissue processing. Generally, frozen sections allow much better antigen preservation than paraffin sections and are favored by some researchers due to the absence of an excessive fixation and processing regime, leaving antigens in a more 'native' form. Markers for estrogen receptors used to fall into this category until antibodies were developed that detected an epitope present in formaldehyde-fixed, paraffin-embedded sections. When prepared correctly, tissue morphology in frozen sections is of a reasonably high quality, but does not appear to compare with that of paraffin sections, probably due to the higher degree of fixation in paraffin sections. Sectioning of frozen tissue is notably more technically demanding, especially if the tissue is calcified or has a high lipid content. To prepare a frozen section, fresh tissue is plunged into a suitable cryo-agent such as liquid nitrogen and once frozen is placed on to a cryostat chuck and embedded in a suitable embedding medium such as OCT (optimal cutting temperature). Once the embedding medium has set, 4 μm sections are obtained by cryotomy and mounted on to a glass microscope slide (see section 2.2.2).

1.3 Free-floating sections

Free-floating sections are commonly favored by those working on sensitive neurological antigens due to the speed at which a result can be obtained. Following cardiac perfusion with an aldehyde-based fixative, tissue is removed from the gross specimen and sectioned, typically at around 50 μm thickness. Sections are then floated out on to a water bath and immunochemically stained *in situ*. The thickness of the specimen allows this to be performed, although particular attention must be paid to washing steps in order to reduce nonspecific background staining due to antibodies becoming 'trapped' in the thick tissue.

1.4 Cytological specimens

Cytological specimens are typically in the form of a conventional smear or as a monolayer obtained via automated technology (for example, Thinprep) or grown on a glass coverslip. Nonspecific background staining can be a problem with cytological specimens, since the clustering of cells can act to 'trap' antibodies. However, this artifact tends to be of less intensity than true positive staining.

1.5 Reproducible and accurate results

It must be stressed that it is the aim of every immunochemical laboratory to obtain reproducible and accurate results every time. There are a plethora of other parameters that have significant effects on the outcome of immunochemical staining. It is common to find scrutiny concerning the reagents and protocols used during an immunochemical staining experiment, for instance reagent concentrations, incubation times, and blocking steps; but in actual fact the results are being predisposed from the point of specimen collection onwards. An equal amount of notice should be paid to how the specimen is treated before being subjected to immunochemical staining as to the staining procedures themselves.

This chapter will discuss the various techniques regarding specimen collection and preparation, followed by the various immunochemical staining methods available to today's researcher, keeping result reproducibility and accuracy in mind at all times. A collective view of quality assurance issues raised in this chapter can be found in Chapter 9.

2. METHODS AND APPROACHES

2.1 Specimen fixation

It is imperative that tissue samples are collected as quickly as possible after the removal of the organ from the subject, or after the subject's death. As soon as the tissue is removed from its *in vivo* environment, it is subjected to influences that are deleterious to antigen preservation such as hypoxia, lysosomal enzymes, and putrefactive changes brought about by bacteria and molds. Adequate fixation of the specimen is necessary to counteract these effects and therefore the time between obtaining the tissue and introducing a fixative should be kept to an absolute minimum. It is for this reason that studies involving neurological antigens often employ cardiac gross specimen perfusion to utilize a quick fixation.

Fixation causes a conformational change in the tertiary structure of the proteins, rendering them biologically inactive. It also creates a marked difference in both the chemical and the antigenic properties of the proteins compared with the native forms. This has a profound significance when considering antibody–antigen interactions, in particular when using a cross-linking fixative (see section 2.1.1). For these reasons, a strict standard operating procedure regarding fixation must be adhered to (see Chapter 9, section 2.1).

Fixatives can be classed simply into those that coagulate proteins and those that exert their preservative role by other mechanisms such as cross-linking (noncoagulative). A more refined way of classifying them is as follows (1):

- Aldehydes: formaldehyde (paraformaldehyde, formalin); glutaraldehyde; acrolein; glyoxal; formaldehyde mixtures containing mercuric chloride, acetic acid, zinc, and periodate lysine.
- Protein-denaturing agents (precipitants): acetic acid; methanol; ethanol; industrial methylated spirits.
- Oxidizing agents: osmium tetroxide; potassium permanganate; potassium dichromate.
- Other cross-linking agents: carbodiimides.
- Physical: heat; microwaves.
- Miscellaneous/unknown: nonaldehyde-containing fixatives, acetone, picric acid.

The most popular fixatives used for immunochemical staining techniques are aldehyde fixatives, protein-denaturing agents, and acetone.

2.1.1 Aldehydes

Of all the known fixatives, the mechanisms of aldehyde fixatives are probably the best understood and will be discussed here in detail. In summary, for immunochemical staining purposes, formaldehyde fixation gives the best compromise between obtaining good cytological preservation and immunolocalization, whilst maintaining antigen masking to a minimum.

Formaldehyde, paraformaldehyde and formalin

Formaldehyde (HCHO) is a low molecular weight gas, the –CHO group being the aldehyde. HCHO molecules readily dissolve in water to form methylene hydrate ($HO\text{-}CH_2\text{-}OH$) and have the same reactivity characteristics as formaldehyde. These methylene hydrate molecules react with each other to form polymers (see *Fig. 1a*).

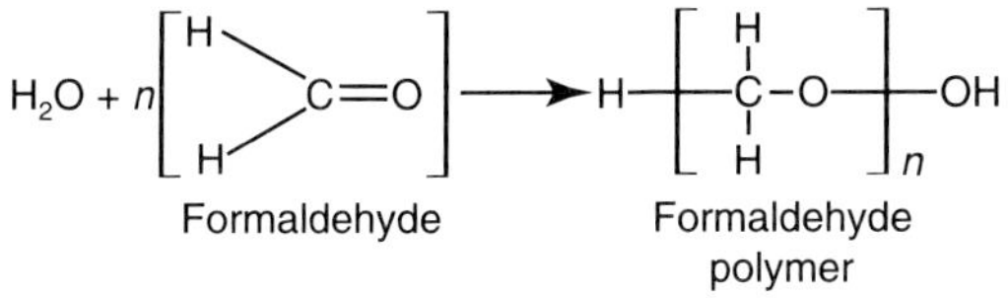

(a) Methylene hydrate molecules react with each other to form polymers

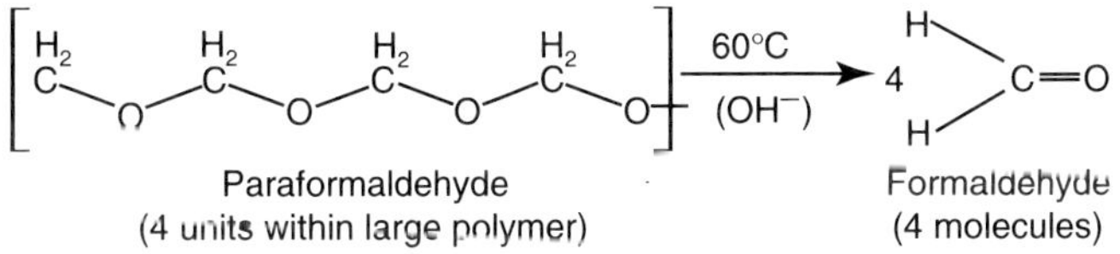

(b) Breakdown of paraformaldehyde to formaldehyde

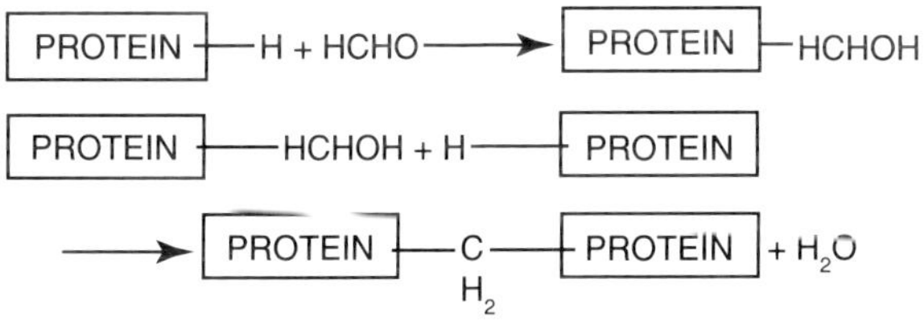

(c) Methylene bridge formation between proteins

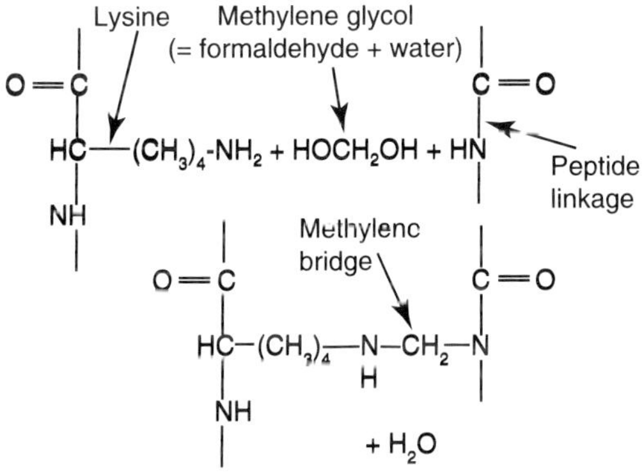

(d) Mechanics of methylene bridge formation

Figure 1. Formation and action of formaldehyde

In this state, formaldehyde consists of 40% (v/v) formaldehyde and 60% (v/v) water and is termed formalin. A typical working formaldehyde solution is diluted to between 4 and 10% (v/v) formaldehyde content. Formaldehyde polymers consist of two to eight repeat units. Polymers of up to 100 repeat units are insoluble and are termed paraformaldehyde.

For adequate fixation to occur, working solutions need to consist predominantly of monomeric methylene hydrate, requiring the breakdown of the polymerized form. Dilution to 10% (v/v) formaldehyde content or below will achieve this, but the reaction occurs passively over several days if plain water is used as the diluent. However, this occurs almost instantaneously when a buffer at physiological pH is used and is further catalyzed by the hydroxide ions in the slightly alkaline solution.

A working solution of formaldehyde can also be generated from paraformaldehyde, with the conversion to monomeric methylene hydrate facilitated by heating to 60°C and the inclusion of a source of hydroxide ions. Richardson (2) used sodium sulfite, but common-day practice is simply to utilize the salts used to buffer the solution to pH 7.2–7.6 (see *Fig. 1b*).

In practical terms, working solutions of formaldehyde show an equal degree of ultrastructural preservation, regardless of whether they are generated from paraformaldehyde or formalin (3). The aldehyde groups form methylene bridge cross-links between adjacent lysine residues on the exterior of the proteins (see *Fig. 1c*). Proteins that were once soluble become anchored to structural proteins and rendered insoluble. The cytoplasm is transformed into a gel-like proteinaceous network, essentially rendering the cell into a condition of stasis that is as close to its *in vivo* state as possible.

The number of cross-links increases with time, and studies conducted on leather tanning suggest that the majority of cross-links occur between the nitrogen atom at the end of the side chain of lysine and the nitrogen atom of a peptide linkage (4) (see *Fig. 1d*).

Other cellular components such as nucleic acids, carbohydrates, and lipids are not fixed directly by formaldehyde, but are trapped in the network of insoluble cross-links. Interestingly, formaldehyde and glutaraldehyde react with DNA and RNA only at 65 and 45°C, respectively (1). This is believed to be related to DNA and RNA uncoiling at these temperatures. With paraffin-embedded tissues, such temperatures are only reached in the tissue processor when wax impregnation occurs, by which time the majority of the available fixative has been washed out.

Tissue hydrophobicity is increased by formaldehyde (and glutaraldehyde) fixation, leading to an increased incidence of general background staining. In addition, specimens fixed in acidic formaldehyde can demonstrate formalin pigment (see Chapter 9, section 2.8.8)

Glutaraldehyde

Glutaraldehyde molecules ($HCO\text{-}(CH_2)_3\text{-}CHO$) are also relatively small, consisting of two aldehyde groups separated by a flexible chain of three methylene bridges (see *Fig. 2*). Glutaraldehyde therefore has double the potential for cross-linking proteins and the flexible CH_2 chain allows this to occur over variable distances.

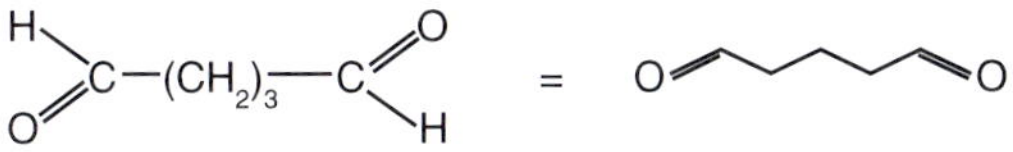

Figure 2. Chemical structure of glutaraldehyde.

Glutaraldehyde exists as polymers of various lengths in aqueous solution (5). Aldehyde groups are present at each end of the polymer, plus one sticking out of each individual polymer unit (see *Fig. 3*). This provides a great ability for binding with lysine molecules, accounting for the rapid fixation of tissue by glutaraldehyde (minutes to hours) compared with that of formaldehyde (see *Fig. 4*).

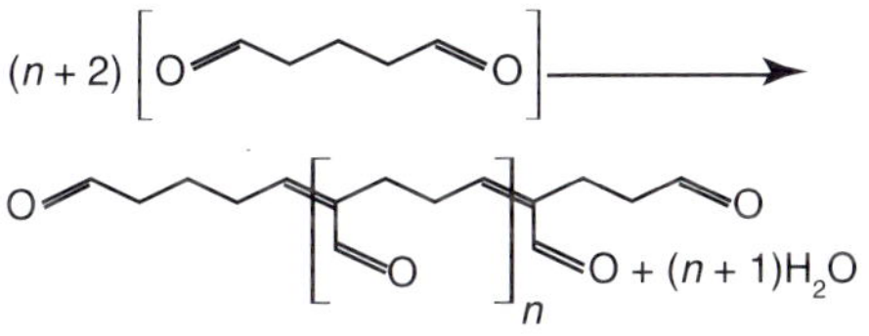

Figure 3. Glutaraldehyde polymer formation.

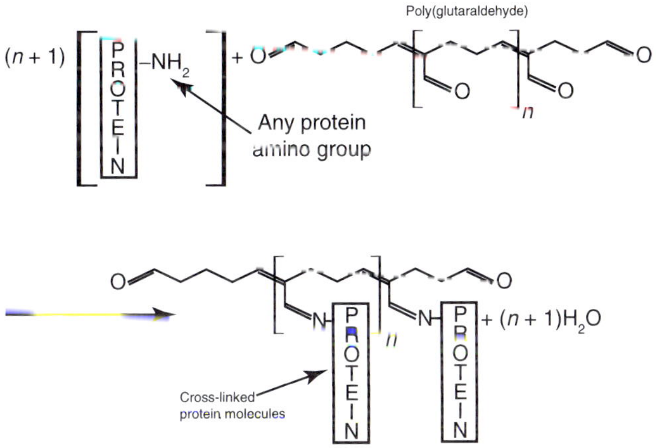

Figure 4. Mechanics of methylene bridge formation.

Free unbound aldehydes are also present that cannot be washed out of the tissue and can provide a source of nonspecific antibody binding if not blocked appropriately (see *Protocol 4*, note d).

As with formaldehyde, glutaraldehyde should ideally be monomeric or at least oligomeric in order to facilitate rapid penetration of tissue. The larger polymers penetrate slowly and in any case the rapid rate of fixation further reduces the rate of penetration. Similarly, tissues fixed in glutaraldehyde are not penetrated well by paraffin wax, making sectioning difficult. However, resin embedding media penetrate glutaraldehyde-fixed tissues well enough to allow the cutting of ultrathin sections for electron microscopy (6).

For immunochemical staining, antigen masking is a major drawback with glutaraldehyde fixation due to the excessive and aggressive cross-linking of proteins. Glutaraldehyde is therefore considered largely unsuitable for tissues destined for immunochemical staining and is best used for electron microscopy where a high degree of cytological preservation is paramount. However, small neurotransmitter molecules such as GABA (γ-aminobutyric acid) can be demonstrated immunochemically in tissues by raising antibodies towards such molecules using glutaraldehyde as the protein carrier linker (see Chapter 2, section 2.2.5). When using such antibodies, inclusion of glutaraldehyde in the formaldehyde fixative is essential, as the epitope comprises the neurotransmitter molecule plus the glutaraldehyde molecule. Fixative mixtures of glutaraldehyde and formaldehyde also give a good compromise, allowing fast stabilization of the tissue by the fast-penetrating formaldehyde, combined with rapid fixation from the slower-penetrating glutaraldehyde. Such mixtures are termed Karnovsky's fixatives and typically contain 1–4% (v/v) glutaraldehyde in 2–4% (v/v) formaldehyde (7).

2.1.2 Protein-denaturing agents

Protein-denaturing agents exert their effect by precipitating proteins. Hydrophobic bonds in the protein's interior are disrupted, changing the protein's tertiary structure. However, secondary structures are maintained since hydrogen bonds are unaffected and appear to be more stable in alcohols than in water. Since protein-denaturing fixatives do not form cross-links between proteins, they are particularly suited for fixing cytological specimens and frozen sections that are destined for immunochemical staining, as no antigen-retrieval step is required. Being friable, such specimens often do not physically survive conventional antigen-retrieval techniques (see section 2.5). The trade-off is that cytological and morphological detail is not as well preserved as with cross-linking fixatives, but this can be beneficial when demonstrating antigens that are sensitive to formaldehyde fixation and tissue processing. If, for whatever reason, formaldehyde is employed to fix frozen or cytological specimens, exposure should be no longer than 5–10 min in order to avoid any high degree of cross-linking.

Ethanol, methanol, and industrial methylated spirits (70% (v/v) ethanol, 30% (v/v) methanol) are commonly used to fix frozen sections, depending largely on the individual laboratory's personal preference and the antigen being demonstrated. Antigens displaying carbohydrate moieties, such as membrane-bound surface antigens, are commonly fixed with alcohol, since alcohol precipitates carbohydrates. However, alcohol fixation tends to hinder cluster designation (CD) marker staining. Frozen sections are usually fixed for 10–30 min in 70–95% (v/v) alcohols in order to help reduce the morphological distortion of nuclear detail and cytoplasm shrinkage seen with absolute alcohol. Some laboratories use an acetic acid:methanol/ethanol mixture for the same reason (8). Acetic acid also aids alcoholic penetration of the tissue.

Alcohol (and acetone) penetrates tissue poorly and generally is only used on tissue sections or cytological preparations rather than pieces of tissue (see Chapter 9, section 2.1.2).

The same protein-denaturing agents used to fix frozen sections can be used to fix cytological preparations, and generally the same principles and considerations apply. There are several commercially available fixative formulations intended especially for cytological specimens, containing alcohol (typically 95% (v/v) ethanol) and polyethylene glycol (PEG). The ethanol fixes the cells and the PEG forms a solid hydrophobic layer over them. This layer protects the cells from mechanical damage and stops the ethanol solution from evaporating, giving adequate cellular fixation and preventing the cells from drying out.

2.1.3 Other fixatives

Acetone

For cell-surface markers such as the CDs, acetone appears to be the most commonly used fixative. However, extended buffer washes during immunocytochemical staining can create undesirable morphological changes in acetone-fixed tissue such as loss of cell membranes and chromatolysis. Chloroform and certain other desiccants have been added to acetone in order to try and reduce these effects, but with no great degree of success (9). Fixing frozen sections at room temperature for 10 min in acetone followed by drying at room temperature for 12–48 h has been reported to improve morphology, but increased background staining may occur. Typically, frozen sections are fixed for 10 min in 100% acetone (after overnight air drying or, if removed directly removed from the freezer, being allowed to warm to room temperature for 10 min) and then immediately immunochemically stained (see section 2.10.2). Some laboratories include an extra 10 min air-drying step after acetone fixation, in order to minimize changes in tissue morphology (see Chapter 9, section 2.1.2). Extended fixation in acetone can make the tissue brittle.

Periodate/lysine/paraformaldehyde

Periodate oxidizes sugars in the tissue to create aldehydes, which are then cross-linked by lysine, whilst the paraformaldehyde cross-links proteins (10). Potassium dichromate is a common additive (termed PLDP) to preserve lipids (11). The three major cellular components are therefore fixed, but the increased degree of cross-linking may accentuate antigen masking.

Bouin's

Caution – picric acid is potentially explosive. Observe appropriate laboratory COSHH (Control Of Substances Hazardous to Health) guidelines.

Bouin's is a formaldehyde-based fixative containing saturated picric acid and glacial acetic acid. Fixation times are similar to those for formaldehyde, although fixation in excess of 24 h can cause some tissues to become brittle. It provides a good degree of preservation for glycogen, but may cause tissues to shrink. Picric acid stains the tissue yellow and after fixation, washes of 50% (v/v) and 70% (v/v) ethanol are used to remove this. Any excess remaining in the tissue sections can be removed with further 70% (v/v) alcohol or 5% (w/v) sodium thiosulfate washes. Bouin's offers excellent morphological preservation and facilitates the removal and chemical alteration of lipids. As this fixative contains formaldehyde, antigen masking may be an issue. Miller (12) claims to have observed reduced immunochemical staining intensity using CD5 (clone 4C7), CD10 (CALLA, clone 56/C6), and cyclin D1 (clone AM29) antibodies on Bouin's-fixed tissues.

B5

B5 is a mercuric chloride/formaldehyde-based fixative. Mercuric chloride is added in order to improve morphological preservation whilst reducing the distortional changes associated with formaldehyde. Mercuric chloride/ formaldehyde-based fixatives do not penetrate tissues well and therefore only small pieces of tissue should be used and fixation periods should be short (2–15 h depending on tissue thickness).

Mercuric chloride/formaldehyde-based fixatives are both cross-linking and coagulative in nature. Their coagulative properties give an exaggerated hardness to tissues, but as these fixatives contain formaldehyde, antigen masking may be an issue. Nuclear and cytoplasmic morphology is well preserved along with good immunolocalization of antigens. Bone marrow and lymph node biopsies are often fixed in B5 for this reason.

Several antigens have been observed to underperform in B5-fixed tissues compared with formaldehyde-fixed tissues. Examples are CD5 (clone 4C7) (13), CD30, and CD23 (12). Antibodies towards kappa and lambda light chains have been reported to perform better in B5-fixed tissues than in formaldehyde-fixed tissues (12).

Sections fixed in B5 may require de-Zenkerization (see below).

Zenker's

Zenker's fixative is also mercuric chloride/formaldehyde-based, but with added potassium dichromate. It essentially performs as for B5, but tissues will require a water wash for 1 h after fixation to remove the potassium dichromate.

Sections fixed in B5 and Zenker's may require de-Zenkerization. This is the term given to removing the mercuric chloride deposits from tissue sections. Sections are treated with 70% (v/v) alcohol containing 0.5% (w/v) iodine for 5 min, washed in running water, decolorized in 5% (w/v) sodium thiosulfate for 2 min, and finally washed in running water again.

De-Zenkerization is often instructed before immunochemical staining, but it has been demonstrated (12, 14) that this should be performed directly before counterstaining, as de-Zenkerization appears to have a deleterious effect on many antigens. The Lugol's iodine has been shown to exert this effect.

Zinc formalin

This fixative gives excellent nuclear morphology and for this reason several authors have claimed it to be preferable to formaldehyde (15–17). As this fixative contains formaldehyde, antigen masking may be an issue.

2.1.4 Other beneficial effects of fixation

In addition to the preservation of protein, fixation can also improve antibody penetration. The gel-like, porous, proteinaceous network of the cytoplasm facilitates the entry of antibodies. The degree of porosity is largely dependent on the type of fixative used, with cross-linking fixatives giving a reduced degree of porosity compared with coagulative fixatives. Noncoagulative fixatives also denature cell membranes by removing lipids, thus further aiding antibody penetration. Acetone alone is particularly efficient at this and some laboratories add nonionic detergents (Triton X-100, Tween 20) or saponin to fixative solutions (18) to solvate lipid membranes. This is not recommended, however, for electron microscopy protocols due to the resulting deterioration in subcellular morphology. Also, in certain instances, detergent permeabilization is detrimental to immunochemical staining, as target proteins are lost from the samples, especially when membrane-bound proteins are being demonstrated. Issues regarding antibody penetration are more applicable to cytological specimens, as the cells have not had the added benefit of being dissected by a microtome blade during sectioning.

Fixation also serves to stabilize the specimen and protect it from the physical rigors of processing and immunochemical staining. Aldehyde fixatives, for example, harden tissue and therefore assist with sectioning and survivability during harsh antigen-retrieval techniques (19).

2.2 Processing

2.2.1 Processing tissues for paraffin wax embedding

Paraffin wax acts as an embedding medium that supports the tissue during microtomy. As fixatives employed for tissues destined for paraffin embedding are aqueous in nature (see section 2.1) and paraffin wax is hydrophobic, there is a need for an intermediate phase that will be miscible with both. A basic tissue-processing regime therefore consists of sending the tissue through graded alcohols of increasing concentration in order to remove the fixative and water, followed by replacement of the alcohol by a 'clearing' solution that is also miscible with the embedding medium, such as xylene (see section 2.7). The embedding medium is then allowed to replace the clearing solution and so obviously needs to be in a liquid form. Histological-grade paraffin wax typically melts between 50 and 60°C, but exposing tissues to this temperature can have deleterious effects on the staining of some antigens, for example vimentin. Tissue sections are also exposed to such temperatures during slide drying, in order to increase their adherence to the slide. The duration and intensity of heating tissues during embedding or slide drying should therefore be kept to a minimum and certainly no longer than overnight.

For further information regarding paraffin processing, see (1).

2.2.2 Processing tissues for frozen sections

Fresh, unfixed tissue destined for cryotomy should be frozen as quickly as possible. This is not only to reduce the deleterious effects on antigen preservation brought about by various post-sampling changes in the tissue (see section 2.1), but also to keep ice crystal formation to an absolute minimum. If frozen slowly, large ice crystals are allowed to form in the tissue, rupturing cell membranes and leading to ice crystal artifacts. Ice crystal artifacts can make morphological interpretation difficult and this has been widely noted in the case of muscle biopsies.

Rapid freezing is commonly achieved by plunging the tissue into liquid nitrogen (–190°C), a technique commonly termed 'snap freezing'. Although rapid freezing is achieved, liquid nitrogen does have its disadvantages. Tissues often shatter due to the rapid expansion of ice and can become too solid, leading to problems with sectioning. It is therefore advisable to allow the tissue to warm to cutting temperature (approx. –20°C) in the cryostat before sectioning. The vapor phase of liquid nitrogen can act as an insulator, thus inhibiting rapid and even freezing of the tissue, further contributing to ice crystal artifacts in certain areas.

The vapor phase issues associated with liquid nitrogen can be overcome by freezing the tissue in isopentane. Isopentane has a higher thermal conductivity (higher freezing temperature, approx. –165°C) than liquid nitrogen. A suitable receptacle containing isopentane is lowered into the liquid nitrogen until the isopentane freezes completely. It is then removed and allowed to warm at room temperature until a partial liquid phase is obtained (approx. –150°C). The tissue is then plunged into the isopentane on a suitable support (such as tin foil) until frozen. Muscle biopsies are frequently frozen in this way.

Whether frozen directly in liquid nitrogen or in liquid nitrogen-cooled isopentane, tissue should be transferred to the interior of the cryostat with maximum haste to minimize any thawing. Once inside the –20°C of the cryostat, any thawed areas of the tissue will refreeze, again contributing to ice crystal formation and artifacts. Handling of the tissue should therefore be kept to a minimum and performed with forceps pre-cooled to the interior temperature of the cryostat and not with warm, gloved fingers! The tissue should be placed on a layer of frozen OCT embedding medium on top of a pre-cooled cryostat chuck. More OCT is then applied around the tissue to give it extra support and allowed to freeze before sectioning.

Frozen sections are then transferred to slides that have been kept at room temperature by gentle but direct contact pressure. As the slides have been kept at room temperature, the contrast in heat helps to 'melt' the frozen sections on to the slide. The slides are then kept within the interior of the cryostat in order for the slides and the sections to re-equilibrate to –20°C or, depending on the laboratory, left to air dry externally. The section's apparent yet brief freeze/thaw cycle does not appear to have any deleterious effects on immunohistochemical staining.

Sometimes it is necessary to perform cryotomy on tissues that have previously been fixed. Fixed tissues have the advantage that fixation helps to reduce the diffusion of labile substances during freezing, such as enzymes and some other antigens. However, the increased water content of the tissues can contribute to a higher degree of ice crystal artifacts. This can be overcome by immersing the tissue into an appropriate sucrose solution for 18 h and blotting dry before snap freezing. As it has a lower freezing temperature than water, the sucrose solution acts as a cryoprotectant and discourages ice crystal formation.

For further information on the production of frozen sections, see (1).

2.3 Specimen storage

Frozen sections can be fixed immediately after cryotomy (commonly in acetone, see section 2.1.3) and immunochemical staining performed, or they can be transferred from the –20°C environment of the cryostat to a –20°C freezer (using a suitably insulated transport vessel such as a polystyrene box containing dry ice) to be used at a later date.

Slides bearing cytological specimens that have been fixed in an alcohol/PEG mixture can be stored at +4°C until required, after the PEG has solidified. In contrast, paraffin-embedded sections can be stored at room temperature for many years without any real noticeable loss of antigenicity (after appropriate antigen-retrieval techniques have been employed (see section 2.5), making storage much easier and more convenient.

2.4 Decalcification

Some specimens require decalcification in order to undergo microtomy to obtain a tissue section. An obvious example is a bone specimen, but other specimens often accumulate calcium deposits as part of pathological processes, such as breast cancer tissue. Calcium deposits in the tissue lead to burring of the microtome blade and subsequent 'scouring' artifacts through the tissue.

Decalcification can be carried out before or after processing to a suitable embedding medium. Specimens that are obviously calcified are decalcified before processing, whilst those that are only discovered to be so following microtomy are surface decalcified. Acid solutions of varying intensities (hydrochloric, nitric, trichloroacetic) are commonly employed as decalcifying agents, as is the chelating agent EDTA.

Decalcification may have a deleterious effect on some antigens (see Chapter 9, section 2.3). A reasonable conclusion to this would be to perform immunochemical staining with decalcification as the only variable and compare results to identify any changes in the staining intensity of the antigen concerned.

For more information on decalcification techniques and the solutions used, see (1)

2.5 Antigen retrieval

The antigen-masking effects of formaldehyde fixation (see section 2.1.1) are reversible to varying degrees by a process known as antigen retrieval. Antigen retrieval involves exposing the tissue sections to heat or proteolytic enzyme digestion before commencing immunochemical staining. The success of retrieval depends on the duration of fixation, the antigen being demonstrated, and the type and conditions of the antigen-retrieval process itself.

Tissues undergoing antigen retrieval should be on tissue-adhesive-coated slides to help adherence to the glass. The harsh conditions of heat-induced epitope (antigen) retrieval (HIER) often cause tissues to lift off the slides, and friable tissues such as breast tissue are prone to lifting, even on coated slides. Slides coated with APES (aminopropyltriethoxysilane) and poly-L-lysine appear to be the most popular and these can be produced in house or purchased ready-made. Of the two, APES-coated slides tend to give the best adherence, especially if an in-house 'double-dipping' procedure is used (see *Appendix 1*).

The mechanisms of antigen retrieval are poorly understood. Both methods are believed to break the methylene bridge cross-links formed between proteins during formaldehyde fixation, thus allowing the proteins to take on a more tertiary-like structure and allowing antibodies access to the epitope. Calcium precipitation (and also that of other divalent metal cations) is currently thought to play a significant role in effective antigen retrieval (20). Methylene bridge formation allows bonding of proteins to calcium ions and their removal or precipitation is thought to be a critical step in salt-mediated antigen retrieval. Morgan (21) demonstrated the chelating effects of EDTA to be improved at high pH compared with low pH.

2.5.1 Proteolytic (enzymatic) antigen retrieval

Proteolytic antigen retrieval can be carried out using a variety of enzyme solutions such as trypsin, pepsin, pronase, and proteinase K. However, it can be said that this method has largely been set aside following the arrival of HIER, first proposed by Shi *et al.* (22), since a much wider range of antigens can be demonstrated (23, 24). Enzymatic antigen-retrieval solutions can also be difficult to work with due to the enzymes being sensitive to factors such as temperature and pH.

2.5.2 Heat-induced epitope (antigen) retrieval

HIER involves heating tissue sections in various buffer solutions and can be performed using a variety of means such as a microwave, pressure cooker, vegetable steamer, or autoclave. Each method has its advantages and disadvantages, but has the common end point of antigen retrieval (see Chapter 9, section 2.4). For instance, domestic microwave ovens are notorious for hot and cold spots, creating an unbalanced retrieval of antigens, and for the rigorous boiling of tissues leading to dissociation from the slide. Scientific models have features such as stirrers and temperature monitoring probes in order to try and alleviate these problems. Buffer solutions commonly used include 0.1 M citrate buffer (pH 6), 0.1 M EDTA (pH 8), 0.5 M Tris base buffer (pH 10), and 0.05 M glycine/HCl buffer.

The factors that have been shown to have the greatest influence on the degree of epitope recovery using HIER are time, temperature, buffer composition, and pH. For instance, it is generally accepted that a difference of 2–3 min in antigen-retrieval time using a pressure cooking can have a noticeable effect on the intensity of immunochemical staining. Shi *et al.* (25) demonstrated that antigens such as L26, PCNA, AE1, EMA, and NSE were retrieved well throughout the pH range of 1–10. Antigens such as MIB1 and ER performed well at very low and neutral to high pH, but extremely well between pH 3 and 6. MT1 and HMB45 showed increased retrieval with increasing pH, but weak demonstration at low pH. No real differences were observed in the degree of antigen retrieval as a result of the buffer used, but Tris/HCl tended to produce better results at high pH than the others. It was concluded that Tris/HCl or sodium acetate buffer at pH 8–9 is suitable for demonstrating most antigens, although some show optimal retrieval at a low pH. Shi *et al.* (26) later observed that Tris/HCl at pH 1 or 10 provided better overall retrieval than citrate buffer at pH 6. The 'cool-down' period after HIER should also be kept constant in order to eliminate this factor as a possible source of result variation.

The demonstration of endogenous biotin levels can be exacerbated after HIER when employing avidin–biotin-based detection systems (see *Protocol 4*) (27). This can be seen in tissues expressing naturally high levels of endogenous biotin, such as liver and kidney, and can lead to erroneous results when perceived as positive staining (28). However, such artifacts can easily be blocked if the appropriate steps are taken during the immunochemical staining procedure (see Chapter 9, section 2.8.6). Negative controls can also be used to help identify the artifact and raises the critical point that negative controls must be exposed to the same antigen-retrieval regime as the test tissue.

There is no such thing as a universal antigen-retrieval solution and method. Each antigen must be taken separately and the variables of temperature, time, buffer composition, and pH optimized accordingly.

2.5.3 Combination of enzyme and heat

Although relatively uncommon, combinations of HIER and proteolytic antigen retrieval may be beneficial in certain cases. Iczkowski *et al.* (29) used such a combination and obtained superior staining for the anti-keratin antibody clone 34βE12 when compared with just one of the techniques alone.

2.5.4 Nonaldehyde-fixed specimens

It has generally been accepted that tissue sections or cytological preparations that have not been fixed in formaldehyde do not require an antigen-retrieval step due to the absence of protein cross-linking. Interestingly, Miller (12) observed often dramatically improved immunochemical staining of several antibodies following pressure cooker HIER of air-dried cytological smears. In contrast, immunochemical staining of alcohol-fixed cytological smears following pressure cooker HIER appeared to be decreased for some antibodies, although all were still detectable. These findings, although unconventional, are worthy of further investigation.

2.6 Counterstaining following immunochemical staining

Few immunochemists appear to give much consideration to nuclear counterstains employed to visualize nuclear detail after immunochemical staining has been performed. However, it is important to get an appropriate contrast between the immunochemical stain and the nuclear counterstain.

Counterstains can be classed into two groups: tinctorial and fluorescent (see *Table 1*). Whichever nuclear counterstain is employed, it is important to use one of a different color to that of the label being used in order to avoid obvious confusion!

Table 1. Common nuclear counterstains

	Color of stained nuclei	Maximum absorption (nm)	Maximum emission (nm)
Tinctorial			
Hematoxylin	Blue	NA	NA
Light green	Green	NA	NA
Fast red	Red	NA	NA
Toluidine blue	Blue	NA	NA
Methylene blue	Blue	NA	NA
Fluorescent			
DAPI	Blue	358	461
Hoechst 33258/33342	Blue	350	450
Propidium iodide	Red	535	617

NA, Not applicable.

2.6.1 Tinctorial stains: hematoxylin

Of all the tinctorial nuclear counterstains, hematoxylin is by far the most popular. Hematoxylin is obtained from the logwood of the tree *Haematoxylon campechianum*. It must be appreciated that the oxidation product of hematoxylin, hematin, is the actual dye and not hematoxylin itself. Oxidation of hematoxylin can be achieved by naturally allowing the process to occur via exposure to light and air, but this can take several months. More commonly, an oxidizing agent is added to the hematoxylin, allowing oxidation to occur almost instantaneously, for example mercuric oxide in the case of Harris's hematoxylin. In comparison, naturally oxidized hematoxylins tend to have a longer shelf-life and give slightly higher-quality staining than force-oxidized hematoxylins, although in practical terms this is negligible.

Hematoxylins can be classified according to how they are mordanted. Hematin itself is anionic and therefore has little ability to bind to chromatin without the presence of a mordant. Mordants for hematoxylin are the salts of iron, aluminum, tungsten, and lead; the former three are more popular than the latter, with aluminum being the most popular overall (see *Table 2*). The mordant exerts a net positive charge on the dye–mordant complex, thus allowing it to bind to and stain the anionic chromatin.

Table 2. Alum-mordanted hematoxylins

Alum hematoxylin	Notes
Harris's	Very popular, commonly used regressively. Beware of alcohol-soluble enzyme/chromogen precipitates
Mayer's	Popular; commonly used progressively
Carazzi's and Gill's	Commonly used progressively

Alum (aluminum-mordanted) hematoxylins can be used regressively or progressively. Progressive hematoxylins are exposed to the tissue/cells for long enough to obtain the desired level of staining before being blued. Regressive hematoxylins are exposed to the tissue/cells for long enough to achieve overstaining and then are differentiated (have some of the stain removed) by immersion in an acid solution (typically 1% acid alcohol (see *Appendix 1*)) for anything up to 1 minute before being blued, depending on the desired degree of differentiation. Hematoxylin initially stains nuclei red until the pH changes to alkaline, when it becomes blue, hence the term 'blueing'. This is commonly achieved during the wash stages following immersion of the tissues/cells in hematoxylin or the differentiation solution. Running tap water in most geographical locations provides enough alkalinity to achieve blueing, but where this is not the case, substitute solutions can be used, such as 0.05% (v/v) ammonia or Scott's tap water (see *Appendix 1*).

Hematoxylins of all varieties can be purchased from various laboratory reagent suppliers and it is interesting to note that the same hematoxylin from different suppliers can often give different results, so it is best to stick to one supplier if possible. Laboratories that produce their own (and often excellent) home-made hematoxylins should pay particular attention to the issues of oxidation and mordanting in order to obtain batch-to-batch consistency (see *Appendix 1*).

Staining intensity depends on several factors, namely the concentration of the hematoxylin solution, the duration of staining, the concentration of the differentiation solution, and the duration of differentiation. The desired level of staining intensity is largely down to personal preference, but there should be a balance between obtaining a good degree of nuclear morphology and not inadvertently masking any nuclear staining.

When using an alcohol-soluble chromogen to visualize antibody binding, such as 3-amino-9-ethylcarbazole (AEC) with a horseradish peroxidase (HRP) label, staining duration in Harris's hematoxylin should be kept to a minimum since it contains alcohol. The use of a nonalcohol-containing hematoxylin such as Mayer's would be more preferable in this situation (see *Table 2*).

Other tinctorial nuclear counterstains can be employed (see *Table 1*), but are not very popular in immunochemical applications in comparison with hematoxylin. Follow the manufacturer's instructions for staining procedures for the relevant stain.

2.6.2 Fluorescent stains

DAPI

DAPI binds selectively to double-stranded DNA with no or very little cytoplasmic staining. Green or red fluorescent labels can easily be used with DAPI, as it does not significantly impair the signal. DAPI can be used on both fixed and unfixed cytological preparations and tissue sections. Follow the manufacturer's instructions for the staining procedure.

Hoechst 33258/33342

Hoechst dyes bind specifically to A/T-rich regions of double-stranded DNA, with no or very little cytoplasmic staining. Their characteristics are very similar to those of DAPI. Follow the manufacturer's instructions for the staining procedure.

Propidium iodide

Propidium iodide intercalates between DNA base pairs with little or no sequence preference and makes a good nuclear counterstain for use with green fluorescent labels. Propidium iodide also binds to RNA, requiring treatment of specimens fixed in aldehyde fixatives with RNase in order to eliminate such staining. Follow the manufacturer's instructions for the staining procedure.

2.7 Mounting following immunochemical staining

Mounting medium is the medium used to make the coverslip (a thin piece of glass) adhere to the tissue section or cytological preparation following immunochemical staining. This is predominantly in order to protect the specimen and the immunochemical staining from physical damage, but also, importantly, to help improve the clarity and contrast of the image during microscopy. Some commercial adhesive mounting media claim to eradicate the need for a coverslip.

The choice of mounting medium following immunochemical staining is largely dictated by the label (and, in the case of enzymatic labels, the chromogen) used to visualize the antigen. Always observe the guidelines given on reagent product inserts.

Mounting media should have the following characteristics (adapted from 8):

1. Be colorless and transparent.
2. Completely permeate and fill the tissue interstices.
3. Have no deleterious effects on the tissue.
4. Resist bacterial contamination.
5. Protect the tissue and staining from mechanical and chemical damage (oxidation and pH changes).
6. Be miscible with the dehydrant or clearing agent.

7. Set without crystallizing, cracking, or shrinking (or otherwise deforming the material being mounted) and not react with, leach, or induce fading in stains and reaction products (including those from enzyme histochemical, hybridization, and immunohistochemical procedures).
8. Once set, the mountant should remain stable (in terms of the features listed above). This is particularly important when long-term specimen storage is required.

Mounting media should ideally have a refractive index (RI) as close as possible to that of the fixed protein (tissue) (approximately 1.53). As light passes from one medium to another, it changes speed and bends. An example of this is the apparent bending of a stick when placed in water. Light travels fastest in a vacuum and in all other media light travels more slowly. The RI of a medium is the ratio of the speed of light in a vacuum to the speed of light in the medium (it is always >1). A mounting medium with an RI close to that of the fixed tissue will therefore render it transparent, with only the stained tissue elements visible. This is where the term 'clearing' comes from – xylene, for instance, has an RI very close to that of fixed tissue, therefore inducing a certain amount of transparency. A mounting medium with an RI too far either side of 1.53 will provide poor clarity and contrast. This can be demonstrated practically by viewing a tissue section with no mounting medium, since air has an RI of 1.0.

Mounting media can be classified into two categories: organic and aqueous (or hydrophobic and hydrophilic, respectively). Organic mounting media can only be used for enzymatic labels where the precipitate formed between the enzyme and the chromogen is not soluble in the alcohols used during dehydration of the tissue (see Chapter 3, section 2.1). The HRP chromogen diaminobenzidine (DAB) is such an example, whereas the HRP chromogen AEC is alcohol soluble. Organic mounting medium is not suitable for fluorescent labels.

Aqueous mounting medium is generally suitable for all enzymatic label/chromogen combinations and fluorescent labels. Specimens mounted in such media are mounted straight from the aqueous phase (with no dehydration or clearing). Once dried, organic mounting media tend to have RIs very close to that of fixed protein. Aqueous mounting media, however, have RIs further away from that of fixed protein than organic media, so the clarity and contrast tends to be poorer in comparison.

Aqueous mounting media for phycobiliprotein fluorescent labels (phycoerythrin, phycocyanin) must not contain glycerol as this quenches the staining intensity. Similarly, exposure to excitation light of most fluorescent labels results in diminished staining, a process known as photobleaching. Fluorescent mounting media commonly contain antifade agents that slow down the photobleaching of such labels, such as 1,4-diazabicyclo[2.2.2]octane (DABCO) and paraphenylenediamine (30), which act as free-radical scavengers. The presence of oxygen and free-radical species is the primary environmental influence on a fluorescent label irreversibly photobleaching.

Both organic and aqueous mounting media can be categorized further as either adhesive or nonadhesive. Adhesive medium hardens due to solvent evaporation, attaching the coverslip firmly to the slide. Nonadhesive medium stays as a liquid and evaporation can be slowed by sealing around the edges of the coverslip with clear nail varnish or paraffin wax. Organic media are adhesive, whilst aqueous media are usually nonadhesive.

From the above, it can be concluded that organic mounting media should always be used over aqueous if the enzymatic label/chromogen combination permits so. For fluorescent labels, coverslips should always be sealed and an antifade agent employed in the media.

The different types of mounting media discussed above can be purchased under many trade names and from numerous suppliers. *Table 3* demonstrates those commonly used.

Table 3. Commonly used organic and aqueous mounting media

Mounting medium	RI	Notes
Organic		
DPX (Distrene, plasticizer, xylene/dibutyl phthalate xylene)	1.523	Adhesive. Very common. Sets quickly
Canada balsam	1.523	Adhesive. Turns yellow with age. Sets slowly
Euparal	1.535	Adhesive. Can fade hematoxylin counterstaining of cell nuclei
Aqueous		
Glycerol	1.46	Nonadhesive. Can be added to other media to slow drying and cracking
90% (v/v) Glycerol (in PBS)	1.47	Nonadhesive. Better RI than glycerol alone. Commonly used for fluorescent labels
Gum arabic + glycerol (Farrant's medium)	1.436	Adhesive. Addition of gum arabic to glycerol solution reduces RI but gives adhesive mountant
Glycerine (glycerol) + gelatin (glycerine jelly)	1.47	Adhesive. Better RI than Farrant's medium
Sucrose + gum arabic (Apathy's medium)	1.5	Adhesive. Better RI than glycerine jelly medium
Polyvinyl alcohol	1.5	Adhesive. Alternative to Apathy's medium. Commonly used for fluorescent labels

2.8 Slide storage following immunochemical staining

The storage conditions of slides following immunochemical staining is largely dictated by the label (and, in the case of enzymatic labels, the chromogen) used to visualize the antigen and the mounting medium. Always observe the storage guidelines given on reagent product inserts.

Fluorescent labels and counterstains that photobleach require storage in the dark. The colored precipitate of the reaction between the enzymatic label HRP and the chromogen AEC will also fade if exposed to sunlight for extended periods. In contrast, the colored precipitate of the reaction between the enzymatic label HRP and the chromogen DAB is not sensitive to sunlight and will keep for many years left on the bench.

Aqueous mounting media of various descriptions will often dry out completely and cause the coverslip to lift and the specimen to dry. Storage at +4°C is often required to help reduce the rate of evaporation, and sealing around the edges of the coverslip will further reduce this. In contrast, organic mounting media such as DPX will dry completely without the coverslip lifting (as long as enough as been applied) and without adversely affecting the specimen, thus being ideal for storage at room temperature. Some mounting media can discolor with age, but this does not usually affect the interpretation of the results.

2.9 Tissue microarrays

The use of tissue microarrays is becoming increasingly popular for both test and quality-control purposes, hence their inclusion in this chapter for completeness.

Conventional tissue slides display around one to ten pieces of tissue per slide, depending on the section size. Tissue microarrays, however, can typically display up to several hundred cores of tissue on a single slide, with the diameter of each core ranging from around 2.0 to 0.6 mm. There are several commercially available technologies for producing such microarrays, taking needle core biopsies from carefully selected areas of donor tissue blocks and forming a grid array in a recipient block. However, areas must be selected carefully to demonstrate the desired tissue antigen.

Tissue microarrays have numerous uses, all intended to increase speed and efficiency and improve standardization:

1. Preparing control tissues. A single microarray could contain one or more cores of control tissue for each antibody routinely used in the laboratory, removing the need for separate control tissues being cut for each antibody. This also has the effect of conserving valuable control

tissues, given the small core diameter compared with the average size of a standard control tissue section. The microarray can be mounted at one end of the slide and the test tissue mounted at the other end, further reducing the number of slides in any one immunochemical staining run. There is also the added benefit that both the test and the control tissues have been subjected to the same immunochemical staining and pre-treatment regimes, thus further reducing errors in experimental variation. However, care should be taken with such slides on automated staining machines that utilize capillary action, as not all of the tissue may be exposed to the antibody if the capillary gap is suboptimal (see Chapter 10, section 2.2.2).

2. Assessing the optimal titer of a new antibody (see Chapter 9, section 2.5.2). A microarray of know positive-control tissue can be prepared and compartmentalized using a hydrophobic barrier pen in order to assess multiple dilutions of the same antibody on a single slide.
3. Assessing the sensitivity and specificity of a new antibody. A microarray of numerous tissues in various states of disease or normality can be prepared to assess the tissue specificity and cellular binding localization of an antibody.
4. High-throughput screening. Tissue microarrays are extremely convenient for high-throughput screening, but concerns can arise regarding the small diameter of each tissue core being nonrepresentative of the entire donor tissue.

2.10 Recommended protocols

The following protocols are robust, tried and tested methods that should provide a solid starting point for obtaining good-quality results when carried out as described. However, each protocol may need optimizing according to the antigen in question and the phenomenon of interlaboratory variation (see section 3).

2.10.1 Antigen-retrieval protocols (see section 2.5)

Protocol 1

Enzymatic method (see section 2.5.1)

Equipment and Reagents

- Water bath containing two troughs (to contain slide racks)
- α-Chymotrypsin (type II from bovine pancreas)

- Calcium chloride
- Ultrapure water
- 0.5% (w/v) Sodium hydroxide solution
- 0.5% (v/v) Hydrochloric acid solution
- Xylene (or other dewaxing reagent)
- 100% Industrial methylated spirits (IMS) or methanol
- Paraffin sections

Method

1. Set the water bath to 37°C. Add the appropriate amount of ultrapure water to each trough and place the troughs into the water bath[a]. Allow the ultrapure water to warm to 37°C.
2. Dewax and rehydrate the paraffin sections by placing them in three changes of xylene for 3 min each, followed by three changes of IMS or methanol for 3 min each, followed by cold running tap water for 3 min.
3. Place the slides in one trough of ultrapure water at 37°C to warm[b].
4. Remove the other trough and into this dissolve 0.1 g of calcium chloride and 0.1 g of chymotrypsin per 100 ml of distilled water, using a magnetic stirrer to ensure that all reagents are properly dissolved[c].
5. Once dissolved, bring the solution to pH 7.8 using the 0.5% (w/v) sodium hydroxide and 0.5% (v/v) hydrochloric acid solutions. Return the trough to the water bath and allow this enzyme solution to reheat to 37°C[d].
6. Transfer the warmed slides into the enzyme solution for a suggested 20 min[e], then remove the slides and place them under cold running tap water for 3 min[f].
7. Continue with an appropriate immunochemical staining protocol (see section 2.10.2).

Notes

[a]Use a sufficient volume of ultrapure water to cover the slides.
[b]Placing cold slides into the enzyme solution will lower the temperature of the solution, thereby reducing enzyme activity and could lead to the antigens being under-retrieved.
[c]Chymotrypsin can be very allergenic. Use a facemask and extraction cabinet for weighing out.
[d]Prepare the chymotrypsin solution as quickly as possible to avoid impairing the activity of the enzyme. Allow this solution to return to 37°C before introducing the slides.
[e]The time period of 20 min is only suggested as a starting point for incubation time. Less than 20 min may leave the antigen under-retrieved, leading to weak staining. More than 20 min may leave the antigen over-retrieved, leading to erroneous staining and also increasing the chances of sections dissociating from the slides. A control experiment is recommended beforehand, where slides of the same tissue section are incubated in the enzyme solution for 10, 15, 20, 25, and 30 min before being immunochemically stained to evaluate the optimum antigen-retrieval time for the particular antigen being demonstrated.
[f]Tap water stops the antigen-retrieval process by washing away the enzyme.

Protocol 2

HIER: pressure cooker method (see section 2.5.2)

Equipment and Reagents

- Domestic stainless steel pressure cooker
- Hotplate
- 2 l Beaker or conical flask
- Trisodium citrate buffer, prepared by mixing: 5.88 g of trisodium citrate, 44 ml of 0.2 M hydrochloric acid solution, and 1956 ml of ultrapure water
- 0.5% (w/v) Sodium hydroxide solution
- 0.5% (v/v) Hydrochloric acid solution
- Xylene (or other dewaxing reagent)
- 100% Industrial methylated spirits (IMS) or methanol

Method

1. Prepare the trisodium citrate buffer by mixing the trisodium citrate, hydrochloric acid, and ultrapure water together in a 2 l beaker or conical flask. Use a magnetic stirrer to ensure that all reagents are properly dissolved.
2. Adjust to pH 6.0 with the 0.5% (w/v) sodium hydroxide and 0.5% (v/v) hydrochloric acid solutions. Add this solution to the pressure cooker. Place the pressure cooker on the hotplate and turn it on to full power. Do not secure the lid of the pressure cooker at this point; simply rest it on top.
3. While waiting for the pressure cooker to come to the boil, dewax and rehydrate the paraffin sections by placing them in three changes of xylene for 3 min each, followed by three changes of IMS or methanol for 3 min each, followed by cold running tap water. Keep them in the tap water until the pressure cooker comes to the boil.
4. Once the pressure cooker is boiling, transfer the slides from the tap water to the pressure cooker. *Take care with the hot solution and steam - use forceps and gloves.* Secure the pressure cooker lid following the manufacturer's instructions.
5. Once the cooker has reached full pressure (see manufacturer's instructions), time for 3 min[a].
6. When 3 min has elapsed, turn off the hotplate and place the pressure cooker in an empty sink. Activate the pressure release valve (see the manufacturer's instructions) and run cold water over the cooker.
7. Once depressurized, open the lid and run cold water into the cooker for 10 min. *Take care with the hot solution and steam.*
8. Continue with an appropriate immunochemical staining protocol (see section 2.10.2).

Note

[a]The time of 3 min is only suggested as a starting point for the antigen-retrieval time. Less than 3 min may leave the antigen under-retrieved, leading to weak staining. More than 3 min may leave the antigen over-retrieved, leading to erroneous staining and also increasing the chances of sections dissociating from the slides. A control experiment is recommended beforehand, where slides of the same tissue section are retrieved for 1, 2, 3, 4, and 5 min before being immunochemically stained to evaluate the optimum antigen-retrieval time for the particular antigen being demonstrated.

Protocol 3

HIER: microwave method

Slides should be placed in a plastic rack for this procedure.

Equipment and Reagents

- Domestic (850 W) or scientific microwave
- 1 l Beaker or conical flask
- Microwaveable vessel, either inbuilt or to hold approximately 400–500 ml
- Trisodium citrate buffer, prepared by mixing: 2.94 g of trisodium citrate, 22 ml of 0.2 M hydrochloric acid solution, and 978.0 ml of ultrapure water
- 0.5% (w/v) Sodium hydroxide solution
- 0.5% (v/v) Hydrochloric acid solution
- Xylene (or other dewaxing reagent)
- 100% Industrial methylated spirits (IMS) or methanol

Method

1. Dewax and rehydrate the paraffin sections by placing them in three changes of xylene for 3 min each, followed by three changes of IMS or methanol for 3 min each, followed by cold running tap water. Keep them in the tap water until step 3.
2. Prepare the trisodium citrate buffer by mixing the hydrochloric acid and ultrapure water together in a 1 l beaker or conical flask. Use a magnetic stirrer to ensure that all reagents are properly dissolved. Adjust to pH 6.0 with the 0.5% (w/v) sodium hydroxide and 0.5% (v/v) hydrochloric acid solutions. Add this solution to the microwaveable vessel[a].
3. Remove the slides from the tap water and place them in the microwaveable vessel. Place the vessel inside the microwave. If using a domestic microwave, set to full power and wait until the solution comes to the boil. Boil for 15 min from this point. If using a scientific microwave, program it so that the antigen is retrieved for 15 min once the temperature has reached 98°C[b].
4. When 15 min has elapsed, remove the vessel and run cold tap water into it for 10 min. *Take care with the hot solution.*
5. Continue with an appropriate immunochemical staining protocol (see section 2.10.2).

Notes

[a]Use a sufficient volume of antigen-retrieval solution to cover the slides. This should be by at least a few centimeters if using a nonsealed vessel to allow for evaporation during the boiling step.
[b]The time of 15 min is only a suggested antigen-retrieval time. Less than 15 min may leave the antigen under-retrieved, leading to weak staining. More than 15 min may leave the antigen over-retrieved, leading to erroneous staining and also increasing the chances of sections dissociating from the slides. A control experiment is recommended beforehand, where slides of the same tissue section are retrieved for 5, 10, 15, 20, 25, and 30 min before being immunochemically stained to evaluate optimum antigen-retrieval time for the particular antibody being used.

2.10.2 Immunochemical staining protocols

Immunochemical staining protocols can vary from simple one-step procedures where the label is conjugated directly to the primary antibody to multiple-step procedures where the label is conjugated to a secondary antibody or an avidin–biotin complex (ABC).

Generally, the more steps that are involved, the more time-consuming and complicated the immunochemical staining technique is to perform (see *Table 4*). However, the trade-off is a greater degree of signal amplification. This is especially important when the antigen in question is present at low concentrations in the specimen. The exceptions to this rule are the polymer (two-step indirect) and ImmPRESS (two-step indirect) methods, providing the second highest sensitivity, but with the quickness and ease of a two-step indirect (nonABC) method.

It is a common misconception when observing immunochemical staining diagrams that only one primary antibody binds to the target, only one secondary antibody binds to the primary antibody, and only one molecule of label or biotin is conjugated to the secondary antibody (see *Fig. 7* in *Protocol 6*). If the primary antibody is monoclonal and the epitope recognized is only present once on the antigen, then indeed only one primary antibody will bind to the antigen. However, if the epitope is present more than once on the antigen, then multiple primary antibodies will bind, provided that molecular hindrance is not an issue. This is especially true for polyclonal antibodies raised using an immunogen that is too large to comprise a single epitope, so the antibodies, by their very nature, will recognize multiple epitopes on the antigen. Secondary antibodies are nearly always polyclonal and so will recognize multiple epitopes on the primary antibody, with several secondary antibodies binding to each of the bound primary antibodies. Each secondary antibody will have several molecules of label or biotin conjugated to it. If an ABC system is being used, then large complexes of avidin–biotin–label will form (see *Fig. 5* in *Protocol 4*), which will bind to the molecules of biotin on the secondary antibodies. The bottom line is that each reagent addition therefore contributes to the overall amplification of the end signal.

The following are general guidelines that are applicable to all immunochemical staining methods:

1. Before proceeding with the following immunochemical staining protocol, ensure that paraffin sections have been dewaxed by processing them through three changes of xylene for 3 min each, followed by three changes of industrial methylated spirits (IMS) or methanol for 3 min each, followed by cold running tap water for 3 min.

Table 4. Pros and cons of common immunochemical staining procedures

Method*	Pros	Cons
One-step direct (nonABC)	Quick and easy to perform	Low sensitivity
Two-step indirect (nonABC)	Higher sensitivity than one-step direct	Slightly more time-consuming to perform than one-step direct, but still relatively quick and easy
Three-step indirect (nonABC)	Higher sensitivity than two-step indirect	Slightly more time-consuming to perform than two-step indirect, but still relatively quick and easy
PAP/APAAP (three-step indirect)	Higher sensitivity than the above techniques as it utilizes the natural affinity of the antibody for the label, rather than a chemical conjugation	Slightly more time-consuming to perform than two-step indirect, but still relatively quick and easy
ABC	Very high sensitivity	Time-consuming and complex to perform
LA/LSA (three-step indirect)	8–10 times more sensitive than the ABC method due to the smaller enzyme–(strept)avidin complex	Time-consuming and complex to perform
ImmPRESS (two-step indirect)	As quick and easy to perform as a two-step indirect (nonABC), with at least the sensitivity of LA/LSA and polymer immunochemical staining techniques, making this method extremely favorable	Few
Polymer (two-step indirect)	As quick and easy to perform as a two-step indirect (nonABC), but with the second highest sensitivity, making this method extremely favorable. Very sensitive (primary antibody dilutions can be up to 20 times greater than for PAP/APAAP immunochemical staining techniques and several times greater than for ABC or LA/LSA immunochemical staining techniques)	Few
Tyramide signal amplification	50 times greater sensitivity than the LA/LSA immunochemical staining technique	Time-consuming and complex to perform

*Shown in order of increasing sensitivity.

Secondly, ensure that any necessary antigen-retrieval techniques have been employed (see section 2.10.1 for the relevant protocols).

2. If using a cytological preparation that has been protected using PEG (see section 2.1.2), remove this by immersion in methanol, ethanol, or IMS for 10 min, followed by a buffer rinse.

3. If using frozen sections, allow them to warm to room temperature for 10 min before fixing them in the appropriate fixative (commonly acetone, see section 2.1.3) for 10 min, followed by a buffer rinse.
4. Perform any necessary antigen-retrieval techniques if aldehyde-fixed specimens are being used (see section 2.5).
5. Carry out all incubations in a humidified chamber so that the sections do not dry out. At no stage during the protocol should drying of the sections be allowed to occur. Drying out will lead to nonspecific binding and ultimately background staining. Commercial chambers are available, but a shallow plastic 'sandwich' box with a sealed lid and wet tissue paper in the bottom will be adequate, as long as the slides can lie flat so that the reagents do not drain off. An excellent homemade solution to the problem of slides not lying flat is to cut a plastic serological pipette into lengths to fit the incubation chamber. Glue these in pairs to the bottom of the chamber, with the two individual pipette tubes of each pair being placed about 4 cm apart. This provides a level and evenly raised surface for the slides to rest on, away from the wet tissue paper.
6. Optimized antibody dilutions should be used to give the highest intensity of positive staining with the lowest amount of background staining. It is advisable to start by using the antibody at the dilution recommended on the supplier's technical datasheet using a known positive-control tissue. In the case of aldehyde-fixed, paraffin-embedded tissue, immunochemically stain one slide with no antigen retrieval, a number of slides with HIER using various buffers and one slide with enzymatic antigen retrieval. Observe the results and determine whether antigen retrieval is necessary and, if so, which method is best. Next, observe the staining pattern. If staining is absent or weak, then the antibody is probably being used at too high a dilution, so repeat the above steps at a lower dilution. If high background is present, then the antibody is probably being used at too low a dilution, so repeat the above steps at a higher dilution. For instance, if the supplier's technical datasheet recommends a dilution of 1:50 and the background is too high, then perform immunochemical staining (using the optimal antigen-retrieval technique) at 1:100, 1:200, and 1:400. For nonaldehyde-fixed tissues, use the same criteria to assess the optimal dilutions but omit the antigen retrieval.
7. Some of the more complex immunochemical staining techniques may be purchased as a commercial kit that includes some or all of the required reagents. If such kits are used, simply insert the commercial reagents into the appropriate section of the protocol.

All of the following protocols assume that an enzymatic label is being used. However, these protocols can easily be adapted for use with fluorescent labels as follows:

1. Carry out all incubations and washes involving the label (either as part of the ABC or directly conjugated to an antibody) in the dark to avoid photobleaching of the label. In the following protocol, this would be from step 9 onwards, assuming that it is the secondary antibody that is labeled. However, if the fluorescent label is directly conjugated to the primary antibody, then all steps from stage 6 must occur in the dark.
2. Use an appropriate fluorescent nuclear counterstain, if desired (see section 2.6).
3. Do not dehydrate and clear before mounting. Use an appropriate aqueous mounting medium and store the slides in the dark to avoid photobleaching (see sections 2.7 and 2.8).
4. Fluorescent staining must be viewed using a fluorescence microscope set up according to the excitation and emission characteristics of the label being used (see Chapter 3, *Table 2*).

Protocol 4

ABC immunochemical staining procedure (see *Fig. 5*)

Equipment and Reagents

Please refer to *Appendix 1* for recipes.

- 4 mM Sodium deoxycholate (only if using cytological preparations – see note b)
- 100% Industrial methylated spirits (IMS) or methanol (only if an enzymatic label is being used and the chromogen allows – see note n)
- ABC in TBS (conjugated to the appropriate label – see note i)
- Chromogen (only if an enzymatic label is being used and if appropriate for the particular type of enzymatic label – see note l)
- Counterstain (appropriate type to contrast with the enzymatic label/chromogen combination or the fluorescent label – see note m)
- Coverslips
- Humidified incubation chamber
- Microscope (appropriate type for the label - see note o)
- Mounting media (appropriate type for the label – see note n)
- Primary antibody optimally diluted in TBS containing 1% (w/v) bovine serum albumin (BSA)
- Secondary biotinylated antibody optimally diluted in TBS containing 1% (w/v) BSA
- TBS
- TBS containing 0.025% (v/v) Triton X-100
- TBS containing 10% (v/v) normal serum and 1% (w/v) BSA
- TBS containing 1% (w/v) BSA
- TBS containing 1.6% (v/v) H_2O_2 (only if using an HRP label – see note j)

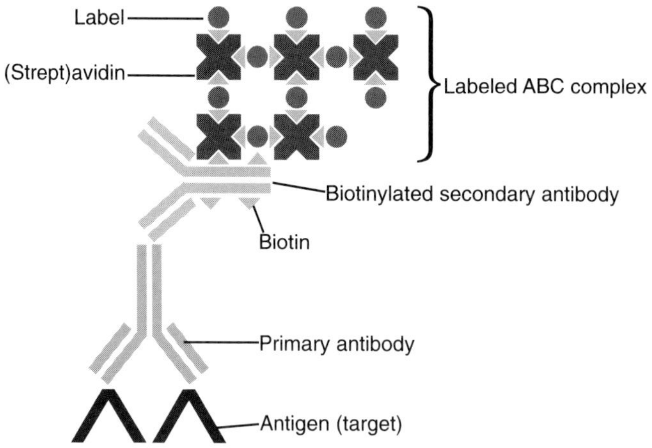

Figure 5. ABC immunochemical staining procedure.

- Xylene (or other dewaxing/clearing reagent) (only for dewaxing paraffin sections and for clearing if an enzymatic label is being used and the chromogen allows – see note n)

Method

Day 1

1. Rinse the slides twice for 5 min each rinse in TBS[a].
2. *For cytological preparations only*: incubate in 4 mM sodium deoxycholate for 10 min[b].
3. Rinse the slides twice for 5 min each rinse in TBS containing 0.025% (v/v) Triton X-100[c].
4. Block in TBS containing 10% (v/v) normal serum and 1% (w/v) BSA for 2 h at room temperature[d].
5. Drain the slides for a few seconds (do not rinse) and wipe around the sections[e].
6. Apply optimally diluted primary antibody in TBS containing 1% (w/v) BSA[f].
7. Incubate overnight at 4°C, preferably on an orbital shaker (gentle agitation)[g].

Day 2

8. Rinse the slides twice for 5 min each rinse in TBS containing 0.025% (v/v) Triton X-100.
9. Apply optimally diluted secondary biotinylated antibody in TBS containing 1% (w/v) BSA for 2 h at room temperature (*at this point make up the ABC following the manufacturer's instructions*)[h,i].
10. Rinse the slides twice for 5 min each rinse in TBS.
11. *If using HRP label only*: incubate in 1.6% (v/v) H_2O_2 in TBS for 30 min at room temperature (5 min for a frozen section or cytological preparation)[j].
12. Rinse the slides twice for 5 min each rinse in TBS.
13. Apply ABC for 30 min at room temperature (made up according to the manufacturer's instructions)[k].

14. Rinse the slides twice for 5 min each rinse in TBS.
15. Develop with the chromogen for 10 min at room temperature (or until the desired degree of staining is achieved as determined under the microscope)[l].
16. Rinse in running tap water for 5 min.
17. Counterstain (if required)[m].
18. Dehydrate, clear, and mount[n].
19. There should now be a colored product localized at the site of antibody binding. This corresponds to the location of the target[o].

Notes

[a]The author recommends TBS to give a cleaner background than PBS. PBS should also be avoided when using an alkaline phosphatase (AP) enzymatic label, as phosphate buffers can quench AP activity. Phosphate is an AP substrate, so will compete with the chromogen.
[b]Sodium deoxycholate is an ionic detergent that permeates the cells by solvating lipid membranes and thus allowing access for the antibody.
[c]The use of 0.025% (v/v) Triton X-100 in the TBS helps to reduce surface tension, allowing the reagents to spread out and cover the whole tissue section with ease. It is also believed to dissolve Fc receptors in frozen sections (see Chapter 9, section 2.8.3 and *Table 1*), thereby helping to reduce specific but undesired background staining. It also serves to improve antibody penetration into the specimen (see section 2.1.4).
[d]Normal serum should be from the species in which the secondary antibody was raised. For instance, if the primary antibody is raised in rabbit and the secondary antibody is a goat anti-rabbit antibody, then normal goat serum should be used. Normal serum helps to prevent the secondary antibody from cross-reacting with endogenous immunoglobulins in the tissue. For example, when detecting an antigen in human tissue and where the secondary antibody is raised in a goat, firstly block using normal goat serum. This will bind to any human immunoglobulins that show cross-reactivity with goat immunoglobulins. On addition of the primary antibody, for example a rabbit antibody of IgG isotype, it will bind to its intended target. On addition of the secondary antibody, for example a biotinylated goat anti-rabbit IgG antibody, it will only bind to the primary rabbit IgG antibody, since if the secondary antibody did have an affinity for any of the endogenous immunoglobulins in the human tissue, it can no longer bind to them as they have already been bound to immunoglobulins in the goat serum. Immunoglobulins from the same species will not interact with each other and therefore the secondary antibody can only bind to the primary antibody.

In addition, antibodies are one of the most hydrophobic of the major serum proteins (which is why they tend to form aggregates when stored over time). The higher a protein's degree of hydrophobicity, the higher its likelihood of linking to another particularly hydrophobic protein; thus hydrophobic interactions between tissue proteins and antibodies can occur. This effect is further exacerbated by aldehyde fixation (see section 2.1.1). Nonspecific binding of the secondary antibody is therefore prevented by pre-immune antibodies in the normal serum binding to and effectively blocking the hydrophobic binding sites in the tissue (since antibodies from the same species will not cross-react). BSA serves the same purpose of reducing nonspecific antibody binding by blocking hydrophobic binding sites in the tissue.

The use of normal serum before the application of the primary antibody also helps to eliminate leucocyte Fc receptor binding of both the primary and secondary antibodies.
[e]Simply remove the excess serum so as not to dilute the primary antibody further. Leaving the serum on the slide when the primary antibody is applied provides additional blocking efficiency.

[f]The primary antibody will target the epitope it is raised against. For proteins, this is an amino acid sequence as unique as practically possible to the protein in question (see Chapter 2, section 1.4). Make sure that the primary antibody is raised in a different species to the tissue being stained. If, for example, mouse tissue is being used and the primary antibody is raised in a mouse, the application of an anti-mouse IgG secondary antibody would bind to all of the endogenous IgG in the mouse tissue and nonspecific background would occur. As discussed in note d, pre-incubation of the tissue with normal mouse serum would have no effect as immunoglobulins from the same species do not interact with each other

[g]Overnight incubation allows antibodies of lower titer or affinity to be used by simply allowing more time for the antibodies to bind. Also, whatever the antibody's titer or affinity for its target, once the tissue has reached saturation point, no more binding can take place. Overnight incubation assures that this occurs. The lower incubation temperature and gentle agitation from an orbital shaker are believed to help reduce background staining by increasing reaction times. *If time is of the essence, apply the primary antibody for only 1 h at room temperature. An overnight incubation is not usually essential, but has the advantages outlined above.*

[h]The secondary antibody recognizes the immunoglobulin species and subtype of the primary antibody. In this example, a biotinylated goat anti-rabbit IgG antibody is being used to bind to the primary rabbit IgG anti-mouse antibody (see note d). The secondary antibody is biotinylated, meaning that it has been conjugated with biotin (see note i).

[i]The ABC consists of biotin–enzyme label conjugates bound to avidin. When applied, the ABC will bind to the secondary biotinylated antibody. After the addition of the secondary antibody, make up the ABC according to the manufacturer's instructions and leave to stand for a minimum of 30 min at room temperature. This is the length of time that the complex takes to form.

Avidin is a protein found in chicken egg white and has similar properties to streptavidin, a protein found in *Streptomyces avidinii.* Both avidin and streptavidin have a high affinity for biotin, a co-factor in enzymes involved in carboxylation reactions. Both avidin and streptavidin can be used in an immunochemical staining application, but streptavidin tends to be favored as it shows greater sensitivity. Streptavidin also produces less nonspecific background staining, due to it being nonglycosylated (unlike avidin), so it shows no interaction with lectins or other carbohydrate-binding proteins.

The ABC is formed through avidin having four binding sites for biotin. The ABC binds to the biotinylated secondary antibody, which is bound to the primary antibody, which is in turn bound to the target on the tissue section.

[j]H_2O_2 suppresses endogenous peroxidase activity and therefore reduces background staining, since endogenous peroxidases as well as the peroxidase label, HRP, would react with the chromogen (see Chapter 3, section 2.3). Using a low concentration for 10 min adequately blocks endogenous peroxidase activity without having a detrimental effect on tissue epitopes. If using AP, then omit this step and step 12. See note l for further details on blocking endogenous AP activity.

It is essential that fresh H_2O_2 is used, as H_2O_2 readily breaks down into water and oxygen at room temperature, rendering it ineffective at blocking endogenous peroxidase activity. H_2O_2 is best stored frozen and thawed shortly before use. AP is therefore an ideal label to use when staining tissue high in endogenous peroxidases, such as spleen.

[k]Due to the lack of BSA, the surface tension of the ABC solution is greater than the other reagents and so will need manually spreading out over the tissue. A 200 μl pipette tip is good for this. Develop the colored product of the enzyme label with the appropriate chromogen. The choice of chromogen depends on the enzyme label being used, the preferred colored end product, and whether aqueous or organic mounting media are being used (see Chapter 3, section 2.1).

[l]Ensure that any chromogen is made up correctly according to the manufacturer's instructions. If using DAB, do not forget that it is a suspected carcinogen. Wear the appropriate protective

clothing. Deactivate it with chloros in a sealed container overnight (it produces noxious fumes when chloros is added) and dispose of it according to laboratory COSHH guidelines. Appropriate COSHH guidelines should be observed regarding the storage, use, handling, and disposal of any laboratory reagents. If using AP, add 0.24 mg/ml levamisole (Sigma L9756) to the chromogen solution. Levamisole suppresses endogenous phosphatase activity and therefore reduces background staining, although not in placenta or the small intestine. The AP label will not be affected by levamisole (see Chapter 3, section 2.1.2).

[m]Apply the appropriate nuclear counterstain, commonly hematoxylin (see section 2.6.1). The desired level of staining intensity is largely down to personal preference, but there should be a balance between obtaining a good degree of nuclear morphology whilst not inadvertently masking any nuclear staining. Try applying the hematoxylin to the specimen for 1 min before differentiation and/or blueing, and adjust accordingly from the results obtained.

[n]If using AEC, Fast Red, INT, or any other aqueous chromogen, note that these are alcohol soluble and a suitable aqueous mounting media should be used. Do not dehydrate and clear before mounting.

Dehydrate and clear DAB, New Fuchsin, Vega Red, NBT, and TNBT (or any other specimens developed using an organic chromogen) by processing them through three changes of methanol (or IMS) for 3 min each, followed by three changes of xylene for 3 min each. Mount the sections in a suitable organic mounting media.

For further information on mounting media, see section 2.7. For further information on labels and chromogens, see Chapter 3.

[o]Enzymatic immunochemical staining should be viewed using a conventional light microscope. Fluorescent immunochemical staining must be viewed using a fluorescence microscope set up according to the excitation and emission characteristics of the label(s) being used (see Chapter 3, *Table 2*).

Protocol 5

One-step direct (nonABC) immunochemical staining procedure (see *Fig. 6*)

Please refer to the relevant notes in *Protocol 4* for each of the steps and to the special considerations for fluorescent labels. *Ensure that the primary antibody is conjugated directly to the label!*

Equipment and Reagents

Please refer to *Appendix 1* for recipes.

- 4 mM Sodium deoxycholate (only if using cytological preparations – see note b)
- 100% Industrial methylated spirits (IMS) or methanol (only if an enzymatic label is being used and the chromogen allows – see note n)
- Chromogen (only if an enzymatic label is being used and appropriate for the type of enzymatic label – see note l)
- Counterstain (appropriate type to contrast with the enzymatic label/chromogen combination or the fluorescent label – see note m)
- Coverslips
- Humidified incubation chamber
- Microscope (appropriate type for the label – see note o)

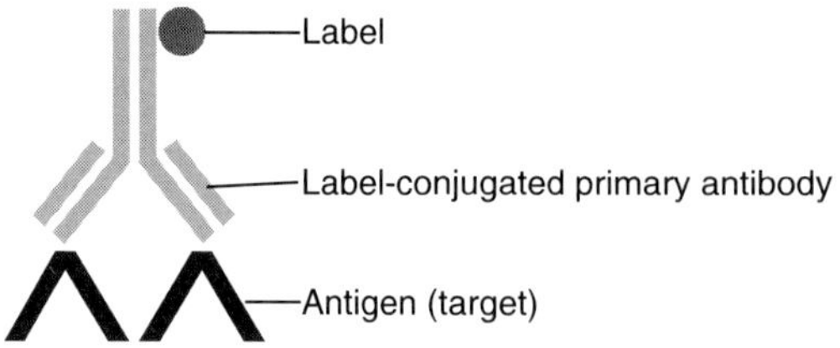

Figure 6. One-step direct (nonABC) immunochemical staining procedure.

- Mounting media (appropriate type for the label – see note n)
- Primary conjugated antibody optimally diluted in TBS containing 1% (w/v) BSA
- TBS
- TBS containing 0.025% (v/v) Triton X-100
- TBS containing 10% (v/v) normal serum and 1% (w/v) BSA
- TBS containing 1% (w/v) BSA
- TBS containing 1.6% (v/v) H_2O_2 (only if using an HRP label– see note j)
- Xylene (or other dewaxing/clearing reagent) (only for dewaxing paraffin sections and for clearing if an enzymatic label is being used and the chromogen allows – see note n)

Method

Day 1

1. Rinse the slides twice for 5 min each rinse in TBS[a].
2. *For cytological preparations only*: incubate in 4 mM sodium deoxycholate for 10 min[b].
3. Rinse the slides twice for 5 min each rinse in TBS containing 0.025% (v/v) Triton X-100[c].
4. *If using HRP label only*: incubate in 1.6% (v/v) H_2O_2 in TBS for 30 min at room temperature (5 min for a frozen section or cytological preparation)[j].
5. Rinse the slides twice for 5 min each rinse in TBS.
6. Block in TBS containing 10% (v/v) normal serum and 1% (w/v) BSA for 2 h at room temperature[d].
7. Drain the slides for a few seconds (do not rinse) and wipe around sections[e].
8. Apply optimally diluted primary conjugated antibody made up in TBS containing 1% (w/v) BSA[f].
9. Incubate overnight at 4°C, preferably on an orbital shaker (gentle agitation)[g].

Day 2

10. Rinse the slides twice for 5 min each rinse in TBS.
11. Develop with chromogen for 10 min at room temperature (or until the desired degree of staining is achieved as determined under the microscope)[l].
12. Rinse in running tap water for 5 min.
13. Counterstain (if required)[m].
14. Dehydrate, clear, and mount[n].
15. There should now be a colored product localized at the site of antibody binding. This corresponds to the location of the target[o].

Protocol 6

Two-step indirect (nonABC) immunochemical staining procedure (see *Fig. 7*)

Please refer to the relevant notes in *Protocol 1* for each of the steps and to the special considerations for fluorescent labels. *Ensure that the secondary antibody is directly conjugated to the label!*

Equipment and Reagents

Please refer to *Appendix 1* for recipes.

- 4 mM Sodium deoxycholate (only if using cytological preparations – see note b)
- 100% Industrial methylated spirits (IMS) or methanol (only if an enzymatic label is being used and the chromogen allows – see note n)
- Chromogen (only if an enzymatic label is being used and appropriate for the type of enzymatic label – see note l)
- Counterstain (appropriate type to contrast with the enzymatic label/chromogen combination or the fluorescent label – see note m)
- Coverslips
- Humidified incubation chamber
- Microscope (appropriate type for the label – see note o)
- Mounting media (appropriate type for the label – see note n)
- Primary antibody optimally diluted in TBS containing 1% (w/v) BSA
- Secondary conjugated antibody optimally diluted in TBS containing 1% (w/v) BSA
- TBS
- TBS containing 0.025% (v/v) Triton X-100
- TBS containing 10% (v/v) normal serum and 1% BSA
- TBS containing 1% (w/v) BSA
- TBS containing 1.6% (v/v) H_2O_2 (only if using an HRP label – see note j)

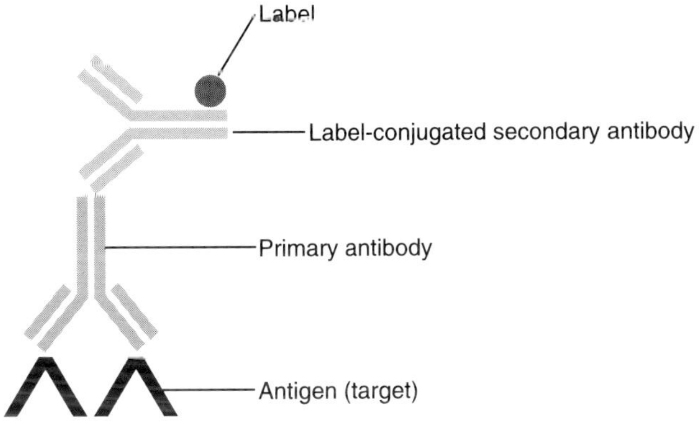

Figure 7. Two-step indirect (nonABC) immunochemical staining procedure.

- Xylene (or other dewaxing/clearing reagent) (only for dewaxing paraffin sections and for clearing if an enzymatic label is being used and the chromogen allows - see note n)

Method

Day 1

1. Rinse the slides twice for 5 min each rinse in TBS[a].
2. *For cytological preparations only*: incubate in 4 mM sodium deoxycholate for 10 min[b].
3. Rinse the slides twice for 5 min each rinse in TBS containing 0.025% (v/v) Triton X-100[c].
4. Block in TBS containing 10% (v/v) normal serum and 1% (w/v) BSA for 2 h at room temperature[d].
5. Drain the slides for a few seconds (do not rinse) and wipe around the sections[e].
6. Apply optimally diluted primary antibody in TBS containing 1% (w/v) BSA[f].
7. Incubate overnight at 4°C, preferably on an orbital shaker (gentle agitation)[g].

Day 2

8. Rinse the slides twice for 5 min each rinse in TBS containing 0.025% (v/v) Triton X-100.
9. *If using HRP label only*: incubate in 1.6% (v/v) H_2O_2 in TBS for 30 min at room temperature (5 min for a frozen section or cytological preparation)[j].
10. Rinse the slides twice for 5 min each rinse in TBS.
11. Apply optimally diluted secondary conjugated antibody made up in TBS containing 1% BSA for 2 h at room temperature[h].
12. Rinse the slides twice for 5 min each rinse in TBS.
13. Develop with chromogen for 10 min at room temperature (or until the desired degree of staining is achieved as determined under the microscope)[l].
14. Rinse in running tap water for 5 min.
15. Counterstain (if required)[m].
16. Dehydrate, clear, and mount[n].
17. There should now be a colored product localized at the site of antibody binding. This corresponds to the location of the target[o].

Protocol 7

Three-step indirect (nonABC) immunochemical staining procedure (see *Fig. 8*)

Please refer to the relevant notes in *Protocol 1* for each of the steps and to the special considerations for fluorescent labels. Also observe the additional notes for this specific

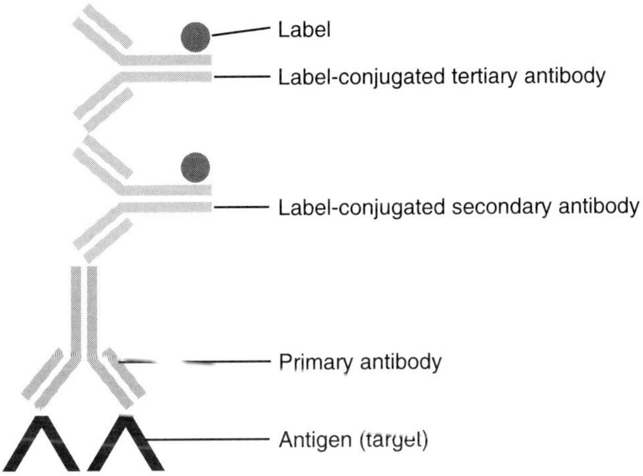

Figure 8. Three-step indirect (nonABC) immunochemical staining procedure.

protocol. *Ensure that the secondary and tertiary antibodies are directly conjugated to the same type of label!*

Equipment and Reagents

Please refer to *Appendix 1* for recipes.

- 4 mM Sodium deoxycholate (only if using cytological preparations – see note b)
- 100% Industrial methylated spirits (IMS) or methanol (only if an enzymatic label is being used and the chromogen allows – see note n)
- Chromogen (only if an enzymatic label is being used and appropriate for the type of enzymatic label – see note l)
- Counterstain (appropriate type to contrast with the enzymatic label/chromogen combination or the fluorescent label – see note m)
- Coverslips
- Humidified incubation chamber
- Microscope (appropriate type for the label – see note o)
- Mounting media (appropriate type for the label – see note n)
- Primary antibody optimally diluted in TBS containing 1% (w/v) BSA
- Secondary conjugated antibody optimally diluted in TBS containing 1% (w/v) BSA
- Tertiary conjugated antibody optimally diluted in TBS containing 1% (w/v) BSA
- TBS
- TBS containing 0.025% (w/v) Triton X-100
- TBS containing 10% (v/v) normal serum and 1% (w/v) BSA (see note p)
- TBS containing 1% (w/v) BSA
- TBS containing 1.6% (v/v) H_2O_2 (only if using an HRP label – see note j)
- Xylene (or other dewaxing/clearing reagent) (only for dewaxing paraffin sections and for clearing if an enzymatic label is being used and the chromogen allows – see note n)

Day 1

1. Rinse the slides twice for 5 min each rinse in TBS[a].
2. *For cytological preparations only*: incubate in 4 mM sodium deoxycholate for 10 min[b].

3. Rinse the slides twice for 5 min each rinse in TBS containing 0.025% (w/v) Triton X-100[c].
4. Block in TBS containing 10% (v/v) normal serum and 1% (w/v) BSA for 2 h at room temperature[d,p].
5 Drain the slides for a few seconds (do not rinse) and wipe around the sections[e].
6. Apply optimally diluted primary antibody in TBS containing 1% (w/v) BSA[f].
7. Incubate overnight at 4°C, preferably on an orbital shaker (gentle agitation)[g].

Day 2

8. Rinse the slides twice for 5 min each rinse in TBS containing 0.025% (w/v) Triton X-100.
9. *If using HRP label only*: incubate in 1.6% (w/v) H_2O_2 in TBS for 30 min at room temperature (5 min for a frozen section or cytological preparation)[j].
10. Rinse the slides twice for 5 min each rinse in TBS.
11. Apply optimally diluted secondary conjugated antibody in TBS containing 1% (w/v) BSA for 2 h at room temperature[h].
12. Rinse the slides twice for 5 min each rinse in TBS.
13. Apply optimally diluted tertiary conjugated antibody in TBS containing 1% (w/v) BSA for 2 h at room temperature[q].
14. Rinse the slides twice for 5 min each rinse in TBS.
15. Develop with chromogen for 10 min at room temperature (or until the desired degree of staining is achieved as determined under the microscope)[l].
16. Rinse in running tap water for 5 min.
17. Counterstain (if required)[m].
18. Dehydrate, clear, and mount[n].
19. There should now be a colored product localized at the site of antibody binding. This corresponds to the location of the target[o].

Additional notes

[p]Apply a mixture of 10% (w/v) normal serum from the species used to raise the secondary antibody, not the tertiary antibody. If a mixture of serums from both of the species used to raise the secondary and tertiary was used, then the tertiary antibody would not only bind to the secondary antibody, but also to any immunoglobulin from the serum of the species of the secondary that is blocking nonspecific binding (see *Protocol 4*, note d).

[q]Ensure that the tertiary antibody is directed towards the immunoglobulin species and subclass of the secondary antibody. Also ensure that the conjugated label is the same as for the secondary antibody.

Protocol 8

Peroxidase anti-peroxidase/alkaline phosphatase anti-alkaline phosphatase (PAP/APAAP) (three-step indirect, nonABC) immunochemical staining procedure (see *Fig. 9*)

Please refer to the relevant notes in *Protocol 1* for each of the steps and to the special considerations for fluorescent labels. Also observe the additional notes for this specific protocol.

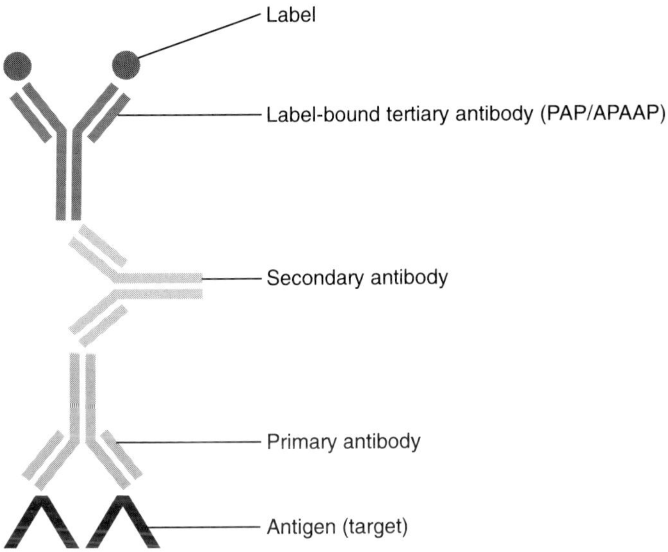

Figure 9. PAP/APAAP (two-step indirect, nonABC) immunochemical staining procedure.

Equipment and Reagents

Please refer to *Appendix 1* for recipes.

- 4 mM Sodium deoxycholate (only if using cytological preparations – see note b)
- 100% Industrial methylated spirits (IMS) or methanol (only if an enzymatic label is being used and the chromogen allows – see note n)
- Chromogen (only if an enzymatic label is being used and appropriate for the type of enzymatic label – see note l)
- Counterstain (appropriate type to contrast with the enzymatic label/chromogen combination or the fluorescent label – see note m)
- Coverslips
- Humidified incubation chamber
- Microscope (appropriate type for the label – see note o)
- Mounting media (appropriate type for the label – see note n)
- Primary antibody optimally diluted in TBS containing 1% (w/v) BSA
- Secondary linker antibody optimally diluted in TBS containing 1% (w/v) BSA
- Tertiary antibody-label complex optimally diluted in TBS containing 1% (w/v) BSA
- TBS
- TBS containing 0.025% (w/v) Triton X-100
- TBS containing 10% (v/v) normal serum and 1% (w/v) BSA
- TBS containing 1% (w/v) BSA
- TBS containing 1.6% (v/v) H_2O_2 (only if using an HRP label – see note j)
- Xylene (or other dewaxing/clearing reagent) (only for dewaxing paraffin sections and for clearing if an enzymatic label is being used and the chromogen allows – see note n)

Method

Day 1

1. Rinse the slides twice for 5 min each rinse in TBS[a].
2. *For cytological preparations only*: incubate in 4 mM sodium deoxycholate for 10 min[b].
3. Rinse the slides twice for 5 min each rinse in TBS containing 0.025% (v/v) Triton X-100[c].
4. Block in TBS containing 10% (v/v) normal serum and 1% (w/v) BSA for 2 h at room temperature[d].
5. Drain the slides for a few seconds (do not rinse) and wipe around the sections[e].
6. Apply optimally diluted primary antibody in TBS containing 1% (w/v) BSA[f].
7. Incubate overnight at 4°C, preferably on an orbital shaker (gentle agitation)[g].

Day 2

8. Rinse the slides twice for 5 min each rinse in TBS containing 0.025% (v/v) Triton X-100.
9. Apply optimally diluted secondary linker antibody in TBS containing 1% (w/v) BSA for 2 h at room temperature[h,r].
10. Rinse the slides twice for 5 min each rinse in TBS.
11. *If using HRP label only*: incubate in 1.6% (v/v) H_2O_2 in TBS for 30 min at room temperature (5 min for a frozen section or cytological preparation)[j].
12. Apply optimally diluted tertiary antibody–label complex in TBS containing 1% (w/v) BSA for 2 h at room temperature[s].
12. Rinse the slides twice for 5 min each rinse in TBS.
15. Develop with chromogen for 10 min at room temperature (or until the desired degree of staining is achieved as determined under the microscope)[l].
16. Rinse in running tap water for 5 min.
17. Counterstain (if required)[m].
18. Dehydrate, clear, and mount[n].
19. There should now be a colored product localized at the site of antibody binding. This corresponds to the location of the target[o].

Additional notes

[r]The secondary antibody acts as a link between the primary antibody and the tertiary antibody–label complex. The primary and tertiary antibodies must therefore be raised in the same species, and the secondary antibody must be raised against the immunoglobulin species and subclass of the primary and secondary antibodies.

[s]Antibodies are raised against the label and then placed in a solution of the label for the antibodies to bind naturally to the label. They are purchased as such a complex.

Protocol 9

Labeled avidin/labeled streptavidin (LA/LSA) (three-step indirect, nonABC) immunochemical staining procedure (see *Fig. 10*)

Please refer to the relevant notes in *Protocol 1* for each of the steps, and to the special considerations for fluorescent labels. Also observe the additional note for this specific protocol.

Equipment and Reagents

Please refer to *Appendix 1* for recipes.

- 4 mM Sodium deoxycholate (only if using cytological preparations – see note b)
- 100% Industrial methylated spirits (IMS) or methanol (only if an enzymatic label is being used and the chromogen allows – see note n)
- LA/LSA complex in TBS (conjugated to the appropriate label – see note t)
- Chromogen (only if an enzymatic label is being used and appropriate for the type of enzymatic label – see note l)
- Counterstain (appropriate type to contrast with the enzymatic label/chromogen combination or the fluorescent label – see note m)
- Coverslips
- Humidified incubation chamber
- Microscope (appropriate type for the label – see note o)
- Mounting media (appropriate type for the label – see note n)
- Primary antibody optimally diluted in TBS containing 1% (w/v) BSA
- Secondary biotinylated antibody optimally diluted in TBS containing 1% (w/v) BSA
- TBS
- TBS containing 0.025% (v/v) Triton X-100
- TBS containing 10% (v/v) normal serum and 1% (w/v) BSA

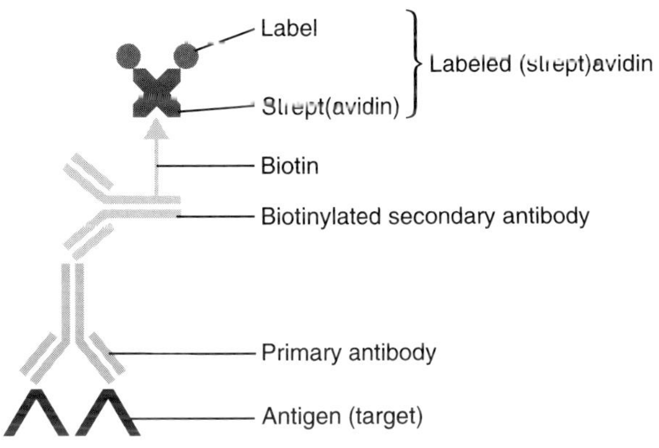

Figure 10. LA/LSA (three-step indirect, nonABC) immunochemical staining procedure.

- TBS containing 1% (w/v) BSA
- TBS containing 1.6% (v/v) H_2O_2 (only if using an HRP label – see note j)
- Xylene (or other dewaxing/clearing reagent) (only for dewaxing paraffin sections and for clearing if an enzymatic label is being used and the chromogen allows – see note n)

Method

Day 1

1. Rinse the slides twice for 5 min each rinse in TBS[a].
2. *For cytological preparations only*: incubate in 4 mM sodium deoxycholate for 10 min[b].
3. Rinse the slides twice for 5 min each rinse in TBS containing 0.025% (v/v) Triton X-100[c].
4. Block in TBS containing 10% (v/v) normal serum and 1% (w/v) BSA for 2 h at room temperature[d].
5. Drain the slides for a few seconds (do not rinse) and wipe around the sections[e].
6. Apply optimally diluted primary antibody in TBS containing 1% (w/v) BSA[f].
7. Incubate overnight at 4°C, preferably on an orbital shaker (gentle agitation)[g].

Day 2

8. Rinse the slides twice for 5 min each rinse in TBS containing 0.025% (v/v) Triton X-100.
9. Apply optimally diluted secondary biotinylated antibody in TBS containing 1% (w/v) BSA for 2 h at room temperature[h].
10. Rinse the slides twice for 5 min each rinse in TBS.
11. *If using HRP label only*: incubate in 1.6% (v/v) H_2O_2 in TBS for 30 min at room temperature (5 min for a frozen section or cytological preparation)[j].
12. Rinse the slides twice for 5 min each rinse in TBS.
13. Apply LA/LSA complex in TBS for 30 min at room temperature (made up according to the manufacturer's instructions)[t].
14. Rinse the slides twice for 5 min each rinse in TBS.
15. Develop with chromogen for 10 min at room temperature (or until the desired degree of staining is achieved as determined under the microscope)[l].
16. Rinse in running tap water for 5 min.
17. Counterstain (if required)[m].
18. Dehydrate, clear, and mount[n].
19. There should now be a colored product localized at the site of antibody binding. This corresponds to the location of the target[o].

Additional note

[t]The biotinylated secondary antibody complexes with the label-conjugated (strept)avidin molecules.

Protocol 10

Polymer (two-step indirect, nonABC) immunochemical staining procedure (see *Fig. 11*)

Please refer to the relevant notes in *Protocol 1* for each of the steps, and to the special considerations for fluorescent labels. Also observe the additional note for this specific protocol.

Equipment and Reagents

Please refer to *Appendix 1* for recipes.

- 4 mM Sodium deoxycholate (only if using cytological preparations – see note b)
- 100% Industrial methylated spirits (IMS) or methanol (only if an enzymatic label is being used and the chromogen allows – see note n)
- Chromogen (only if an enzymatic label is being used and appropriate for the type of enzymatic label – see note l)
- Counterstain (appropriate type to contrast with the enzymatic label/chromogen combination or the fluorescent label – see note m)
- Coverslips
- Humidified incubation chamber
- Microscope (appropriate type for the label – see note o)
- Mounting media (appropriate type for the label – see note n)
- Primary antibody optimally diluted in TBS containing 1% (w/v) BSA
- Secondary polymer complex (made up according to the manufacturer's instructions)
- TBS
- TBS containing 0.025% (v/v) Triton X-100
- TBS containing 10% (v/v) normal serum and 1% (w/v) BSA
- TBS containing 1% (w/v) BSA
- TBS containing 1.6% (v/v) H_2O_2 (only if using an HRP label – see note j)
- Xylene (or other dewaxing/clearing reagent) (only for dewaxing paraffin sections and for clearing if an enzymatic label is being used and the chromogen allows – see note n)

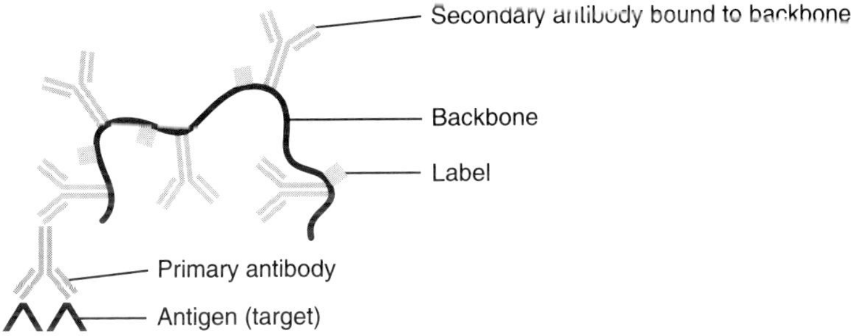

Figure 11. Polymer (two-step indirect, nonABC) immunochemical staining procedure.

Method

Day 1

1. Rinse the slides twice for 5 min each rinse in TBS[a].
2. *For cytological preparations only*: incubate in 4 mM sodium deoxycholate for 10 min[b].
3. Rinse the slides twice for 5 min each rinse in TBS containing 0.025% (v/v) Triton X-100[c].
4. Block in TBS containing 10% (v/v) normal serum and 1% (w/v) BSA for 2 h at room temperature[d].
5. Drain the slides for a few seconds (do not rinse) and wipe around the sections[e].
6. Apply optimally diluted primary antibody in TBS containing 1% (w/v) BSA[f].
7. Incubate overnight at 4°C, preferably on an orbital shaker (gentle agitation)[g].

Day 2

8. Rinse the slides twice for 5 min each rinse in TBS containing 0.025% (v/v) Triton X-100.
9. *If using HRP label only*: incubate in 1.6% (v/v) H_2O_2 in TBS for 30 min at room temperature (5 min for a frozen section or cytological preparation)[j].
10. Rinse the slides twice for 5 min each rinse in TBS.
11. Apply secondary polymer complex for 1 h at room temperature (made up according to the manufacturer's instructions)[u].
12. Rinse the slides twice for 5 min each rinse in TBS.
13. Develop with chromogen for 10 min at room temperature (or until the desired degree of staining is achieved as determined under the microscope)[l].
14. Rinse in running tap water for 5 min.
15. Counterstain (if required)[m].
16. Dehydrate, clear, and mount[n].
17. There should now be a colored product localized at the site of antibody binding. This corresponds to the location of the target[o].

Additional note

[u]This technology utilizes immunoenzyme labels attached to a dextran polymer backbone, conjugated to secondary antibodies. Such polymers are usually supplied ready-to-use. Polymers do not contain avidin or biotin, making them an excellent choice for high-sensitivity staining in tissues containing high levels of endogenous biotin. There is potential for the large polymer backbone to hinder staining due to steric hindrance of the antibody for its target, but in practical terms this is probably negligible and is counteracted by the high degree of amplification.

Protocol 11

ImmPRESS (two-step indirect, nonABC) immunochemical staining procedure (see *Fig. 12*)

Please refer to the relevant notes in *Protocol 1* for each of the steps, and to the special considerations for fluorescent labels. However, at the time of going to press, only HRP conjugation is currently available. Also observe the additional note for this specific protocol.

Equipment and Reagents

Please refer to *Appendix 1* for recipes.

- 4 mM Sodium deoxycholate (only if using cytological preparations – see note b)
- 100% Industrial methylated spirits (IMS) or methanol (only if an enzymatic label is being used and the chromogen allows – see note n)
- Chromogen (only if an enzymatic label is being used, and appropriate for the type of enzymatic label – see note l)
- Counterstain (appropriate type to contrast with the enzymatic label/chromogen combination or the fluorescent label – see note m)
- Coverslips
- Humidified incubation chamber
- Microscope (appropriate type for the label - see note o)
- Mounting media (appropriate type for the label – see note n)
- Primary antibody optimally diluted in TBS containing 1% (w/v) BSA
- ImmPRESS secondary reagent (supplied ready to use; Vector Laboratories)
- TBS
- TBS containing 0.025% (v/v) Triton X-100
- TBS containing 10% (v/v) normal serum and 1% BSA
- TBS containing 1% (w/v) BSA
- TBS containing 1.6% (v/v) H_2O_2 (only if using an HRP label – see note j)
- Xylene (or other dewaxing/clearing reagent) (only for dewaxing paraffin sections and for clearing if an enzymatic label is being used and the chromogen allows – see note n)

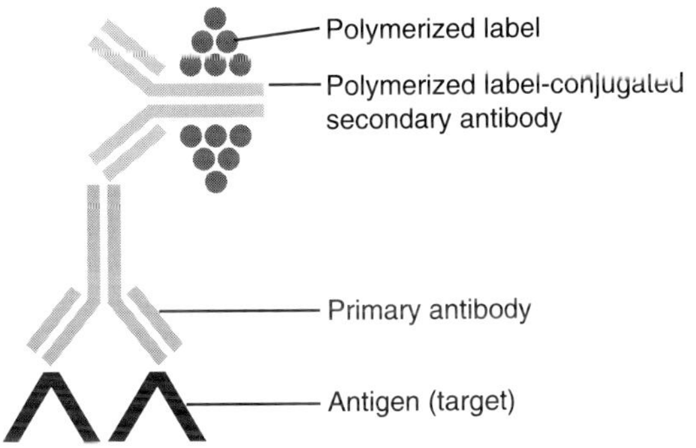

Figure 12. ImmPRESS (two-step indirect, nonABC) immunochemical staining procedure.

Method

Day 1

1. Rinse the slides twice for 5 min each rinse in TBS[a].
2. *For cytological preparations only*: incubate in 4 mM sodium deoxycholate for 10 min[b].
3. Rinse the slides twice for 5 min each rinse in TBS containing 0.025% (v/v) Triton X-100[c].
4. Block in TBS containing 10% (v/v) normal serum and 1% (w/v) BSA for 2 h at room temperature[d].
5. Drain the slides for a few seconds (do not rinse) and wipe around the sections[e].
6. Apply optimally diluted primary antibody in TBS containing 1% (w/v) BSA[f].
7. Incubate overnight at 4°C, preferably on an orbital shaker (gentle agitation)[g].

Day 2

8. Rinse the slides twice for 5 min each rinse in TBS containing 0.025% (v/v) Triton X-100.
9. *If using HRP label only*: incubate in 1.6% (v/v) H_2O_2 in TBS for 30 min at room temperature (5 min for a frozen section or cytological preparation)[j].
10. Rinse the slides twice for 5 min each rinse in TBS.
11. Apply ready-to-use ImmPRESS secondary reagent for 30 min at room temperature[v].
12. Rinse the slides twice for 5 min each rinse in TBS.
13. Develop with chromogen for 10 min at room temperature (or until the desired degree of staining is achieved as determined under the microscope)[l].
14. Rinse in running tap water for 5 min.
15. Counterstain (if required)[m].
16. Dehydrate, clear, and mount[n].
17. There should now be a colored product localized at the site of antibody binding. This corresponds to the location of the target[o].

Additional note

[v]Multiple molecules of polymerized immunolabels are directly conjugated to the secondary antibodies, hence avoiding potential steric hindrance problems that can be associated with polymer backbones (see *Protocol 7*). Potential nonspecific binding is also claimed to be reduced, due to the absence of the polymer backbone and better diffusion of the reagents to the target sites.

Protocol 12

Tyramide signal amplification immunochemical staining procedure (see *Fig. 13*)

Please refer to the relevant notes in *Protocol 1* for each of the steps, and to the special considerations for fluorescent labels. Also observe the additional notes for this specific protocol. This immunochemical staining technique is basically an ABC method with an added signal amplification step.

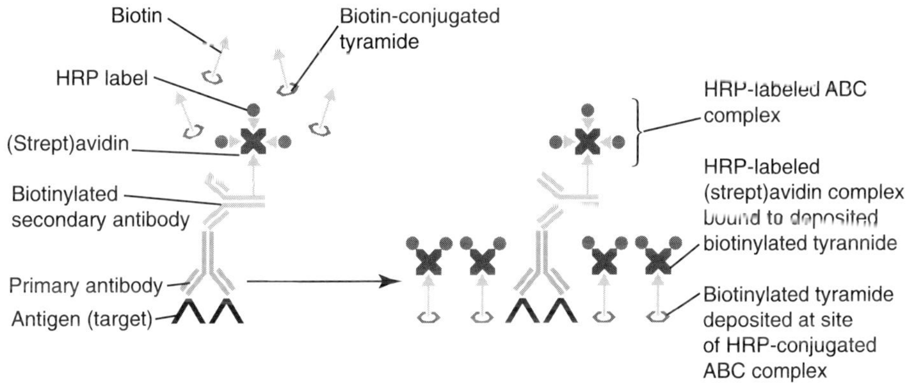

Figure 13. Tyramide signal amplification immunochemical staining procedure.

Equipment and Reagents

Please refer to *Appendix 1* for recipes.

- 4 mM Sodium deoxycholate (only if using cytological preparations – see note b)
- 100% Industrial methylated spirits (IMS) or methanol (only if an enzymatic label is being used and the chromogen allows – see note n)
- ABC in TBS (conjugated to the appropriate label – see note l)
- Biotinyl-tyramide amplification reagent (made up according to the manufacturer's instructions; NEN Life Science Products)
- Chromogen (for HRP label)
- Counterstain (appropriate type to contrast with the enzymatic label/chromogen combination or the fluorescent label – see note m)
- Coverslips
- HRP-conjugated LA/LSA reagent (made up according to the manufacturer's instructions)
- Humidified incubation chamber
- Microscope (appropriate type for the label – see note o)
- Mounting media (appropriate type for the label – see note n)
- Primary antibody optimally diluted in TBS containing 1% (w/v) BSA
- Secondary biotinylated antibody optimally diluted in TBS containing 1% (w/v) BSA
- TBS
- TBS containing 0.025% (v/v) Triton X-100

- TBS containing 10% (v/v) normal serum and 1% (w/v) BSA
- TBS containing 1% (w/v) BSA
- TBS containing 1.6% (v/v) H_2O_2 (only if using an HRP label – see note j)
- Xylene (or other dewaxing/clearing reagent) (only for dewaxing paraffin sections and for clearing if an enzymatic label is being used and the chromogen allows – see note n)

Method

Day 1

1. Rinse the slides twice for 5 min each rinse in TBS[a].
2. *For cytological preparations only*: incubate in 4 mM sodium deoxycholate for 10 min[b].
3. Rinse the slides twice for 5 min each rinse in TBS containing 0.025% (v/v) Triton X-100[c].
4. Block in TBS containing 10% (v/v) normal serum and 1% (w/v) BSA for 2 h at room temperature[d].
5. Drain the slides for a few seconds (do not rinse) and wipe around the sections[e].
6. Apply optimally diluted primary antibody in TBS containing 1% (w/v) BSA[f].
7. Incubate overnight at 4°C, preferably on an orbital shaker (gentle agitation)[g].

Day 2

8. Rinse the slides twice for 5 min each rinse in TBS containing 0.025% (v/v) Triton X-100.
9. Apply optimally diluted secondary biotinylated antibody in TBS containing 1% (w/v) BSA for 2 h at room temperature (*at this point make up the ABC following the manufacturer's instructions*)[h,i].
10. Rinse the slides twice for 5 min each rinse in TBS.
11. Incubate in 1.6% (v/v) H_2O_2 in TBS for 30 min at room temperature (5 min for a frozen section or cytological preparation)[j].
12. Rinse the slides twice for 5 min each rinse in TBS.
13. Apply ABC in TBS for 30 min at room temperature[k].
14. Rinse the slides twice for 5 min each rinse in TBS.
15. Apply the biotinyl-tyramide amplification reagent for 30 min at room temperature (made up according to the manufacturer's instructions)[w].
16. Rinse the slides twice for 5 min each rinse in TBS.
17. Apply HRP-conjugated LA/LSA reagent for 30 min at room temperature (made up according to the manufacturer's instructions)[x].
18. Rinse the slides twice for 5 min each rinse in TBS.
19. Develop with chromogen for 10 min at room temperature[l].
20. Rinse in running tap water for 5 min.
21. Counterstain (if required)[m].
22. Dehydrate, clear, and mount[n].
23. There should now be a colored product localized at the site of antibody binding. This corresponds to the location of the target[o].

Additional notes

[w]The HRP labels of the ABC act as catalysts to increase the number of biotinylated phenol molecules precipitated around the binding site of the primary antibody.

[x]The HRP-conjugated LA/LSA then binds to the precipitated biotinylated phenol molecules to further increase the number of HRP molecules around the binding site of the primary antibody. The overall result is massive signal amplification capable of detecting minute concentrations of antigen.

3. TROUBLESHOOTING

Troubleshooting immunochemical staining procedures is not an easy task. With so many variables, it is often difficult to know exactly where to begin. The best approach is to start with the simple and progress to the more technically demanding issues later. *The importance of using appropriate controls cannot be overstressed.*

Please see Chapter 9, section 3, for further information on controls and troubleshooting.

4. REFERENCES

★★ 1. **Bancroft J & Stevens A (eds.)** (1996) *Theory and Practice of Histological Techniques*, 4th edn. Churchill Livingstone, NY, USA.

2. **Richardson KC** (1960) *J. Anatomy*, **94**, 457–472.

3. **Carson FL, Martin JH & Lynn JA** (1973) *Am. J. Clin. Pathol.* **59**, 365–373.

4. **Gustavson KH** (1956). *The Chemistry of Tanning Processes.* Academic Press, NY, USA.

5. **Monsan P, Puzo G & Marzarguil H** (1975) *Biochimie* **57**, 1281–1292.

6. **Horobin RW & Tomlinson A** (1976) *J. Microsc.* **108**, 69–78.

7. **Karnovsky MJ** (1965) *J. Cell Biol.* **27**, 137A–138A.

★★ 8. **Woods AE & Ellis RC** (1994) *Laboratory Histopathology: a Complete Reference.* Churchill Livingstone, Edinburgh, UK.

9. **Stead RH, Bacolini M & Leskovec M** (1985) *Can. J. Med. Technol.* **47**, 162–178.

10. **McLean IW & Nakane PK** (1974) *J. Histochem. Cytochem.* **22**, 1077–1083.

11. **Holgate CS, Jackson P, Pollard K, Lunny D & Bird CC** (1986) *J. Pathol.* **149**, 293–300.

★★★ 12. **Miller RT** (2001) *Technical Immunohistochemistry: Achieving Reliability and Reproducibility of Immunostains.* Society for Applied Immunohistochemistry Annual Meeting, 8 September, 2001, NY, USA.

13. **Dorfman DM & Shahsafaei A** (1997) *Mod. Pathol.* **10**, 859–863.

14. **Facchetti F, Alebardi O & Vermi W** (2000) *Am. J. Surg. Pathol.* **24**, 320–322.

15. **Abbodonzo SL, Allred DC, Lampkin S & Bank PM** (1991) *Arch. Pathol. Lab. Med.* **115**, 31–33.

16. **Arnol MM, Srivastava S, Fredenburgh J, Stockard CR, Myers RB & Grizzle WE** (1996) *Biotech. Histochem.* **71**, 224–230.

17. **Dapson RW** (1993) *Biotech. Histochem.* **68**, 75–82

18. **Laurila P, Virtanen I, Wartiovaara J & Stenman S** (1978) *J. Histochem. Cytochem.* **26**, 251–257.
19. **Hopwood, D** (1969) *Histochem. J.* **1**, 323–360.
20. **Morgan JM, Navabi H, Schmid KW & Jasani B** (1994) *J. Pathol.* **174**, 301–307.
21. **Morgan JM, Navabi H & Jasani B** (1997) *J. Pathol.* **182**, 233–237.
22. **Shi S-R, Key ME & Kalra KL** (1991) J. *Histochem. Cytochem.* **39**, 741–748.
23. **Gown AM, de Wever N & Battifora H** (1993) *Appl. Immunohistochem.* **1**, 256–266.
24. **Leong AS-Y & Milios J** (1993) *Appl. Immunohistochem.* **1**, 267–274.
25. **Shi S-R, Imam A, Young RJ, Cote RJ & Taylor CR** (1995) *J. Histochem. Cytochem.* **43**, 193–201.
26. **Shi S-R, Cote RJ, Chaiwun B, *et al.*** (1998) *Appl. Immunohistochem.* **6**, 89–96.
27. **Rodriguez-Soto J, Warnke RA & Rouse RV** (1997) *Appl. Immunohistochem.* **5**, 59–62.
28. **Iezzoni JC, Mills SE, Pelkey TJ & Stohler MH** (1999) *Am. J. Clin. Pathol.* **111**, 229–234.
29. **Iczkowski KA, Cheng L, Crawford BG & Bostwick DG** (1999) *Mod. Pathol.* **12**, 1–4.
30. **Balatan AJ, Dalix AM & Oriol R** (1985) *Arch. Pathol. Lab. Med.* **109**, 108.

CHAPTER 5
Multiple immunochemical staining

Ian William Jones and Adam Westmacott

1. INTRODUCTION

Multiple immunochemical staining facilitates the visualization of two or more proteins of interest concomitantly within a sample and, as such, is an invaluable histological resource. The ability to localize multiple proteins within the same preparation permits detailed analyses of the spatial relationships between tissue components and allows protein distributions to be mapped in context with specific cellular and subcellular markers. There are numerous possible multiple immunochemical staining methods available to choose from, utilizing a range of different labels (1, 2). This chapter outlines the major common strategies for multiple immunochemical staining at the light, fluorescence, and electron microscope levels. The methods described are adaptable and may be used with a range of tissue preparations, labels, and microscopy techniques (see *Fig. 1*, available in the color section, page xiv).

Multiple immunochemical staining is an extension of single immunochemical staining (see Chapter 4, section 2.10.2) and the underlying principles are basically the same. An appreciation and understanding of these basics is essential for the design and implementation of robust multiple immunochemical staining protocols and will aid in troubleshooting any problems that arise during the course of the experiments.

1.1 Choosing an appropriate method

Successful multiple staining relies on the ability to distinguish individual immunoreactivities within a sample, and a range of methodologies exist to achieve this aim. The procedures differ in the manner in which the individual immunoreactivities are visualized, some being suitable for light and/or electron microscopy and others for fluorescence/confocal microscopy. Choosing the most appropriate method for a particular combination of antibodies and target proteins will greatly enhance the chances of

Immunohistochemistry: *Methods Express* (S. Renshaw, ed.)

successfully identifying individual staining patterns in a single sample. The advantages and disadvantages of each method are discussed in section 2.

A simple guide to choosing suitable immunochemical staining protocols for different microscopy techniques is shown in *Fig. 2*. Whilst any of the particular methods within a group may be utilized, some may be preferable to others under certain situations. For instance, immunoenzyme techniques (see *Protocol 1*) may not be suitable for visualizing target proteins localized within the same subcellular compartment, as the overlapping chromogens can be difficult to distinguish. In this instance, multiple immunofluorescence (see *Protocol 2*) or immunogold + immunoenzyme (*Protocol 4*) may be more appropriate strategies, as overlapping immunoreactivities are easier to visualize.

1.2 Experimental design

Before undertaking any multiple immunochemical staining experiment, it is strongly advised that the staining strategy is planned out carefully in advance. By taking time to go through the sequential immunochemical staining steps, the chances of success are increased. To aid in designing a multiple staining experiment, it is often useful to sketch a diagram of the complete staining strategy (see *Fig. 3*) as this helps to visualize the

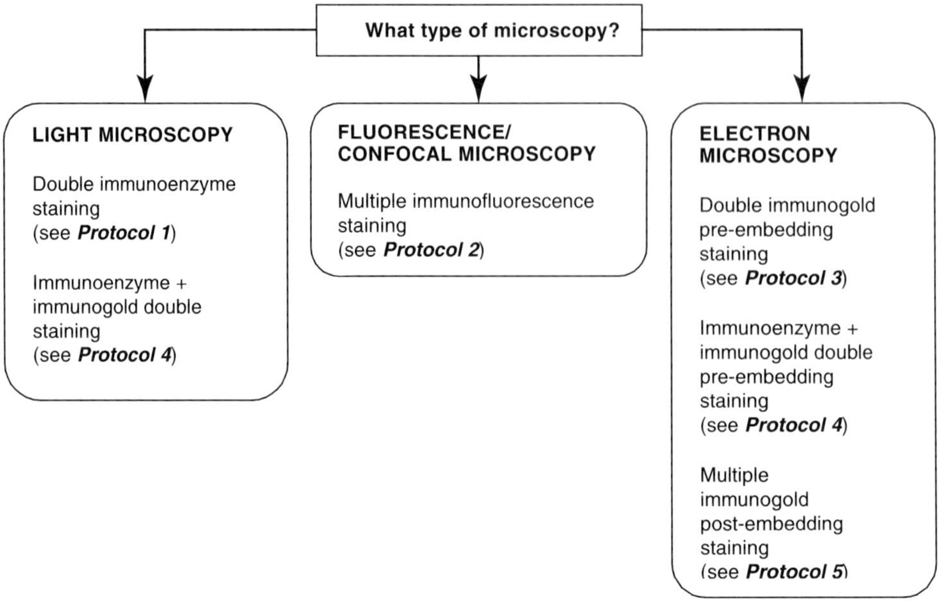

Figure 2. Choosing an appropriate multiple immunochemical staining technique.

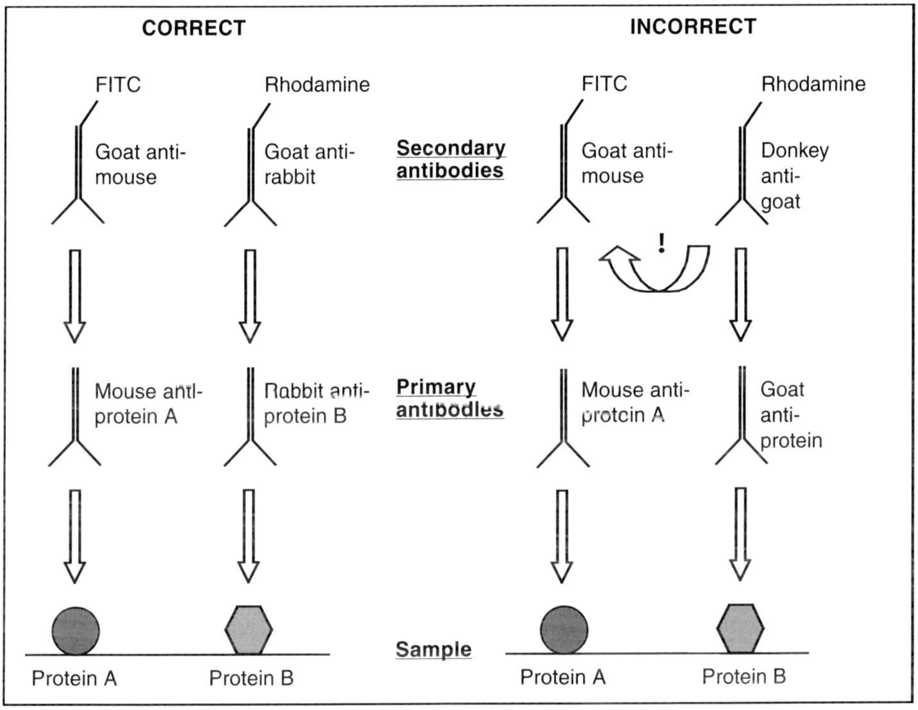

Figure 3. Designing a multiple immunochemical staining experiment.

interactions between individual elements of the staining process and will identify any particular problems that may arise. For example, *Fig. 3* highlights two possible strategies to co-localize two proteins (A and B) in the same sample using the double immunofluorescence technique (see *Protocol 2*). The scheme on the left is correctly designed, as there is no cross-reactivity between the secondary immunoreagents used to visualize the two proteins because it uses secondary antibodies from the same species (goat). In each column, the arrows point straight down, indicating that there will be no cross-reactions between immunoreagents. The scheme on the right, however, is poorly designed as the donkey anti-goat secondary antibody used to visualize protein B will also bind to the goat anti-mouse secondary antibody used to visualize protein A (indicated by the arrow crossing between columns). This cross-reactivity would lead to erroneous instances of co-staining and hence false co-localization.

In addition to designing a robust staining strategy, there are several other aspects of the immunochemical staining process that should be optimized to achieve reliable multiple staining:

1.2.1 Fixation and tissue processing

Not all antibodies work under all fixation regimes. Likewise, the manner in which the tissue is processed (frozen sections, paraffin wax-embedded sections, etc.) will also affect the antigenicity of the target proteins. Therefore, it is critical that a fixation regime and tissue preparation method is chosen that are compatible with all of the primary antibodies to be utilized in the multiple staining experiment. Consequently, any multiple staining studies should be preceded by a series of single staining experiments for each target protein to identify compatible fixation and microtomy techniques. These initial experiments will also provide immunoreactivity benchmarks for each of the target proteins and will help in determining which multiple staining method is most suitable (see *Fig. 2*; also see Chapter 4, sections 2.1 and 2.2).

1.2.2 Secondary immunoreagents

There is a wide range of commercially available secondary immunoreagents and labels to reveal primary antibody binding sites in tissue preparations (see Chapter 3). Whilst most of these products are compatible with multiple staining experiments, careful choice of the most applicable can improve the quality of staining significantly and make subsequent analyses easier.

Secondary antibodies

Using highly pre-absorbed secondary antibodies, in which antibodies that cross-react with nontarget species antibodies have been removed, will reduce the chances of unwanted cross-reactions between immunoreagents. These purified antibodies are particularly useful when the primary antibody species are closely related, such as mouse and rat. In this instance, it is recommended to use anti-mouse secondary antibodies that have been pre-absorbed against rat immunoglobulin and anti-rat secondary antibodies that have been pre-absorbed against mouse immunoglobulin. These pre-absorbed antibodies tend to be more expensive than nonabsorbed antibodies, but are a worthwhile investment and are strongly recommended (see Chapter 2, section 2.2.1).

Immunoenzyme substrates (chromogens)

When used in multiple staining experiments, chromogen substrates should be easily distinguished from each other. There are numerous commercially available substrates available for each of the commonly used enzymes used in immunochemical staining studies, such as horseradish peroxidase (HRP), alkaline phosphatase (AP) and glucose oxidase (see Chapter 3, section 2.1). One common combination used with peroxidase-conjugated immunoreagents is diaminobenzidine (DAB) (brown) and nickel–DAB (black; see *Protocol 1*). Be careful to ensure that the substrates used are compatible with the mounting media to be used, as some are incompatible with organic media and will leach from the samples, resulting in a loss of the label (see Chapter 3, section 2.5).

Fluorochromes

As with immunoenzyme substrates, there is a bewildering array of fluorochromes to choose from, varying in their excitation and emission properties, photostability, and relative brightness (see Chapter 3, section 2.2). Selection of fluorochromes is entirely dependent on the available filter sets in the fluorescence or confocal microscope to be used in image analysis (see Chapter 6, section 2.2). Where possible, fluorochromes with overlapping excitation and emission wavelengths should be avoided to reduce the possibility of one fluorochrome exciting a different fluorochrome. This is less of an issue with newer confocal microscope systems, as many of these are capable of separating emission wavelengths of closely related fluorochromes (see Chapter 6, section 2.9).

1.3 Appropriate controls

Appropriate positive and negative controls are vital to any immunochemical staining experiment. Without these reference markers, it is difficult to analyze immunochemical staining patterns critically with any certainty. There are numerous control conditions that can be performed, such as peptide pre-absorption of polyclonal primary antibodies, omission of primary antibodies, and omission of secondary immunoreagents. The number of possible control combinations is increased in multiple staining studies due to the use of several parallel immunodetection pathways and can be extremely daunting. Ideally, all of the possible controls should be undertaken to scrutinize immunoreagent and staining specificity rigorously. In reality, time constraints and sample availability often restrict the number of control conditions that can be tested. In these cases, the following core controls should be performed: (i) Positive controls: single staining for each of the target proteins using the appropriate label; and (ii) negative controls: omission of one of the primary antibodies but retaining all of the secondary immunoreagents. Therefore, for a double staining experiment (as in *Fig. 3*), there would be four control conditions:

1. Single immunochemical staining for protein A.
2. Single immunochemical staining for protein B.

3. Multiple immunochemical staining, omitting anti-protein A primary antibody.
4. Multiple immunochemical staining, omitting anti-protein B primary antibody.

An additional control, which can be of use when there are doubts over the immunochemical staining patterns, is to swap over the labels and see if the pattern is repeated (see Chapter 9, section 2.6).

1.4 Multiple staining using same-species primary antibodies

One of the most common obstacles encountered in the design of multiple staining experiments is the proposed use of primary antibodies from the same host species. This issue arises frequently as, at present, a majority of monoclonal and polyclonal antibodies are raised in mice and rabbits, respectively. Consequently, it is often necessary to use two or more mouse or rabbit primary antibodies in the same preparation. The difficulty stems from the fact that the secondary antibodies are unlikely to be able to discriminate between the different same-species primary antibodies, leading to cross-reactivity and erroneous co-localization of labels. Similar problems can also be experienced when using primary antibodies from closely related species, such as goat and sheep, which share common epitopes on their immunoglobulins. To circumvent these problems, there are a number of possible adaptations that facilitate the use of same-species primary antibodies in multiple immunochemical staining studies.

1.4.1 Using directly labeled primary antibodies

The simplest method for co-immunochemical staining with primary antibodies from the same species is to incubate samples with primary antibodies directly conjugated to labels, such as enzymes, fluorochromes, or gold particles. This strategy alleviates the need for secondary antibody incubations, thereby avoiding completely the possibility of immunoreagent cross-reactivity. Whilst being the most basic approach, this method does have several drawbacks.

Firstly, it can be difficult to source suitable label-conjugated primary antibodies. Many antibody suppliers do sell appropriate conjugates of their most popular primary antibodies, but the range is limited and it may not be possible to obtain the desired product. An alternative is to conjugate the antibodies in house by means of commercially available, easy-to-use kits. Such kits are widely available, enabling direct conjugation of a variety of labels to both primary and secondary antibodies. This latter approach obviously has an added cost value, but it may be an attractive option for frequently used antibodies.

Secondly, directly conjugated primary antibodies may have restricted use if the target protein is in low abundance in the sample. Secondary immunoreagents amplify staining intensity through the binding of several antibodies to a primary antibody and the presence of several labels on each secondary antibody (see Chapter 4, section 2.10.2). This signal amplification is lost when secondary immunoreagents are omitted from the protocol, resulting in a significant reduction in signal intensity. To overcome this problem, primary antibodies can be conjugated to molecules that are themselves amenable to signal enhancement. For instance, biotinylated primary antibodies can be visualized using avidin–biotin complex (ABC) label conjugates. Such conjugates are widely available, as are kits for in-house biotinylation of primary antibodies (see Chapter 4, section 2.10.2, *Protocol 4*).

1.4.2 Using class- and subclass-specific secondary antibodies

This is the preferred method when immunochemical staining with multiple monoclonal primary antibodies of differing immunoglobulin classes, such as IgG and IgM (see Chapter 2, section 1). These monoclonal antibodies can be readily distinguished using anti-IgG and anti-IgM secondary immunoreagents that are widely available. Likewise, monoclonal antibodies of defined IgG subclass, such as IgG1 and IgG2a, can be labeled separately using subclass-specific secondary antibodies. This method is not compatible with polyclonal primary antibodies, as these tend to contain a mixture of immunoglobulin classes and subclasses.

1.4.3 Using conjugated Fab fragments for blocking and staining

Monovalent Fab fragment, formed from the cleavage of immunoglobulins, can be utilized in a variety of ways to facilitate multiple labeling using same-species primary antibodies. Unlike whole immunoglobulins and $F(ab)_2$ fragments, monovalent Fab fragments only contain one antigen-binding site, so are less likely to cross-link immunoreagents, thereby reducing the likelihood of cross-reactivity (see Chapter 2, section 1.4). As Fab fragments are typically directed against specific immunoglobulin classes, such as IgG and IgM, their use is not recommended when the primary antibodies are of different immunoglobulin classes or subclasses. In these cases, class-specific or subclass-specific antibodies should be used to distinguish between the two primary antibodies (see section 1.4.2).

Label-conjugated Fab fragments can be used in a single step to label a primary antibody and, at the same time, block binding of subsequent immunoreagents (3) (see *Protocol 6* and *Fig. 4*). This method may, however, require a high concentration of conjugated Fab fragments to achieve effective blocking of the first primary antibody, which could result in a high background. If this occurs, lowering the concentration of the conjugated Fab and subsequent blocking with unconjugated Fab may further improve the signal-to-noise ratio. Additionally, small aggregates of conjugated Fab may be present, which could act as divalent or polyvalent molecules, capturing some of the second primary antibody and resulting in overlapping detection of antigens. In this situation, one of the following two methods may be more applicable.

1.4.4 Using *unconjugated* Fab fragments to convert the first primary antibody into a different species

In this instance, unconjugated Fab fragments are utilized to switch the effective species of one of the primary antibodies, permitting the use of a standard multiple staining protocol designed to detect primary antibodies

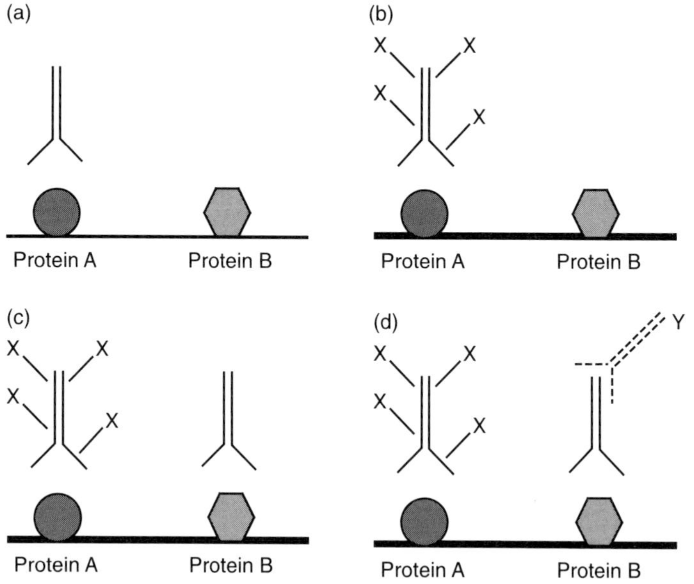

Figure 4. Multiple staining using same-species primary antibodies: the use of conjugated Fab fragments for blocking and staining.
(*a*) Incubate with the first primary antibody, against protein A. (*b*) Incubate with Fab fragments, conjugated to label X. (*c*) Incubate with the second primary antibody, against protein B. (*d*) Incubate with secondary antibody, conjugated to label Y.

of differing species (4) (see *Protocol 7* and *Fig. 5*). In this scenario, it is important to verify that the tertiary antibody used to detect the Fab fragments does not recognize the host species of either the primary antibodies or the second, secondary antibody.

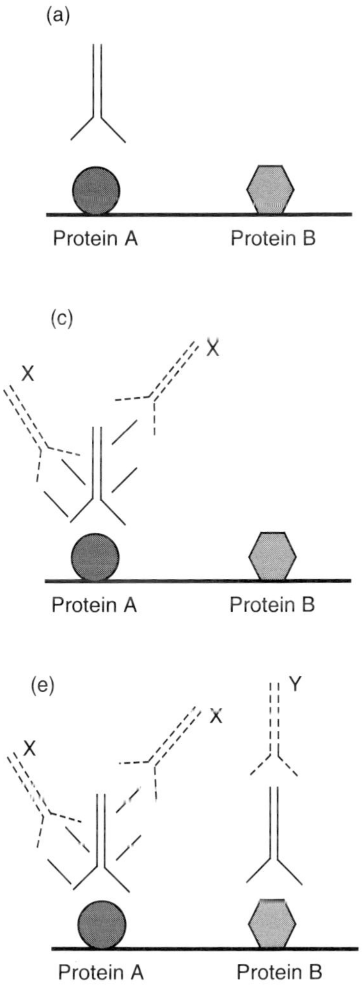

Figure 5. Multiple staining using same-species primary antibodies: the use of unconjugated Fab fragments to convert the first primary antibody into a different species.
(*a*) Incubate with the first primary antibody, directed against protein A. (*b*) Incubate with unconjugated Fab fragments. (*c*) Incubate with secondary antibody conjugated to label X and directed against species of Fab fragments. (*d*) Incubate with the second primary antibody, directed against protein B. (*e*) Incubate with secondary antibody, conjugated to label Y.

1.4.5 Using *unconjugated* Fab fragments to block after the first secondary antibody step

In this method, individual proteins are labeled sequentially, using Fab fragments to saturate any free antigen-binding sites after the first immunoreaction (5, 6) (see *Protocol 8* and *Fig. 6*). This is a more complicated

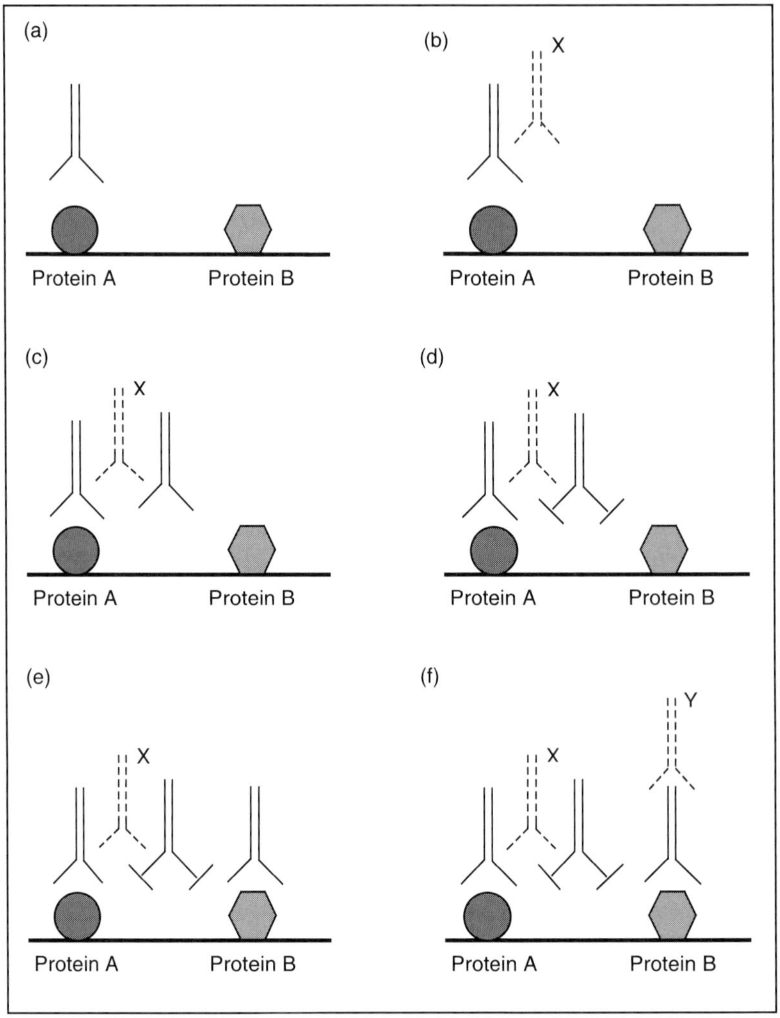

Figure 6. Multiple staining using same-species primary antibodies: the use of unconjugated Fab fragments to block after the first secondary antibody step. (*a*) Incubate with the first primary antibody, directed against protein A. (*b*) Incubate with secondary antibody conjugated to label X. (*c*) Incubate with normal serum from the same host species as the primary antibody. The normal serum acts as a source of nonimmune IgG, saturating any open antigen-binding sites on the first secondary antibody so that it cannot bind the second primary antibody. (*d*) Incubate with unconjugated Fab fragments directed against primary antibody species. (*e*) Incubate with the second primary antibody, directed against protein B. (*f*) Incubate with secondary antibody, conjugated to label Y.

and time-consuming method than the previous two examples but has the advantage of employing unconjugated Fab fragments, which can be used with a variety of different labels.

2. METHODS AND APPROACHES

As illustrated in *Fig. 2*, there are several possible strategies for immunolocalizing two or more proteins of interest within the same sample. The first step in choosing a multiple staining protocol is to decide which microscopy technique is going to be used. This will help determine which labels can be used to localize the individual target proteins within the sample. Light and electron microscopy techniques are compatible with a range of immunoenzyme and immunogold labels, and these can be used in a variety of combinations, such as double immunoenzyme, double immunogold or immunoenzyme + immunogold. Likewise, there is a range of fluorescent labels that can be used in fluorescence and confocal microscopy studies (see Chapter 3, *Table 2*). An understanding of the advantages and disadvantages of each class of label will greatly assist in identifying the most appropriate combinations for a given multiple staining study.

2.1 Labels for light microscopy techniques

The most common type of label used in light microscopy immunochemical staining is immunoenzyme substrates, such as DAB. There is a range of immunoenzymes available, the most common being HRP, AP, and glucose oxidase (see Chapter 3, *Table 1*). Multiple staining can be performed using the same class of immunoenzyme, as in double immunoperoxidase, or using combinations of immunoenzymes, such as immunoperoxidase + immunophosphatase (7) (see *Fig. 7A*, also available in the color section). There is a wide choice of colored substrates for each class of enzyme. For multiple immunoenzyme staining studies, combinations of enzyme substrates should be chosen that are easy to distinguish from each other within the same sample. It is also important to ensure that the substrates used are all compatible with the mounting media used at the end of the protocol.

For some protein targets, such as membrane-bound receptors, it may be preferable to use immunogold particles for visualization of bound antibodies. Unlike immunoenzyme substrates that can diffuse within subcellular compartments, immunogold particles provide more precise information on the subcellular distribution of proteins, as they remain in

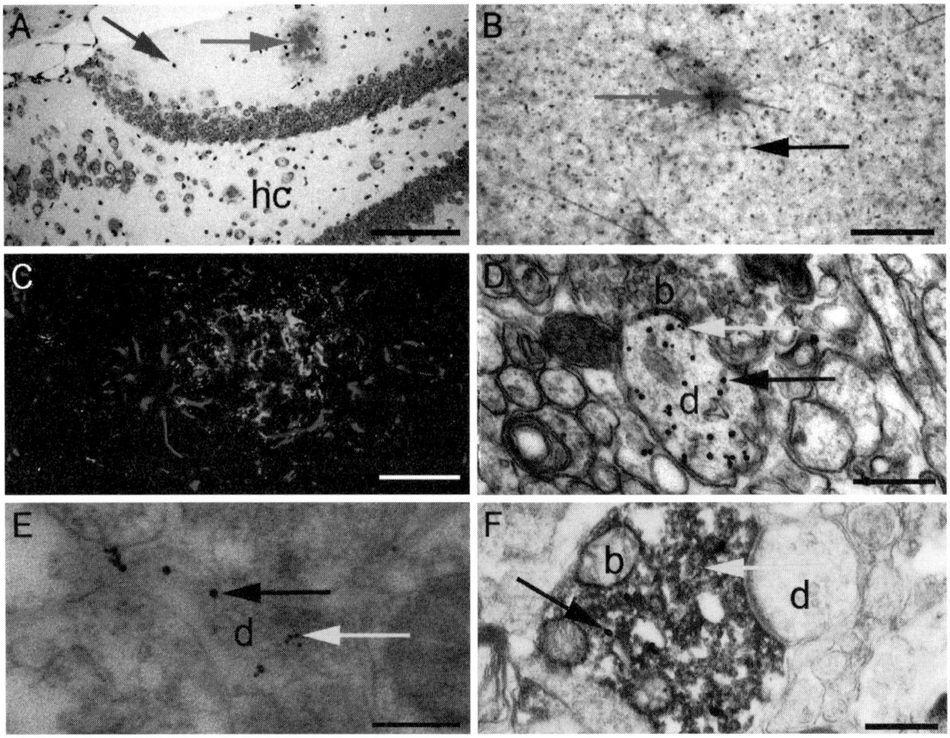

Figure 7. Examples of multiple staining at the light and electron microscope levels (see page xv for color version).
(*A*) Multiple immunoenzyme staining. Wax-embedded section through the mouse hippocampus dual labeled for glial fibrillary acidic protein (Vector VIP peroxidase substrate; red arrow) and β-amyloid (Vector Red alkaline phosphatase substrate; blue arrow) using *Protocol 1*. The section is counterstained with Nissl to reveal neuronal cell bodies (blue). (*B*) Immunoenzyme + immunogold dual staining for light microscopy. Wax-embedded section through the mouse hippocampus dual labeled for GFAP (Vector Red alkaline phosphatase substrate; red arrow) and α7 neuronal nicotinic acetylcholine receptors (nanogold-conjugated α-bungarotoxin; black arrow) using *Protocol 4*. (*C*) Multiple immunofluorescence staining. Frozen section through the mouse hippocampus dual labeled for GFAP (Alexa Fluor 546-conjugated secondary antibody; red) and β-amyloid (Alexa Fluor 488-conjugated secondary antibody; green) using *Protocol 2*. (*D*) Double immunogold pre-embedding staining. Vibrating microtome section through the mouse midbrain dual labeled for vesicular glutamate transporters (large gold particles; black arrow) and α7 neuronal nicotinic acetylcholine receptors (small gold particles; green arrow) using *Protocol 3*. (*E*) Double immunogold post-embedding staining. Acrylate resin section through mouse midbrain dual labeled for tyrosine hydroxylase (large gold particles; black arrow) and β2 neuronal nicotinic acetylcholine receptor subunits (small gold particles; green arrow) using *Protocol 5*. (*F*) Immunoenzyme + immunogold dual staining for electron microscopy. Vibrating microtome section through the mouse midbrain dual labeled for α7 neuronal nicotinic acetylcholine receptors (gold particles; black arrow) and vesicular glutamate transporters (DAB peroxidase substrate; green arrow) using *Protocol 4*. b, Axonal bouton; d, dendrite; hc, hippocampus. Bars, 200 mm (*A*), 20 mm (*B*), 10 mm (*C*), and 200 nm (*D–F*).

close proximity to their targets. Immunogold particles are also amenable to semi-quantitative analysis as the discrete particles can be readily visualized, enabling statistical evaluation of protein expression (8, 9). Although such analyses are possible using enzyme substrates, they are difficult to control as staining intensity is proportional to substrate incubation time and can vary between experiments. For most experiments, the smallest possible immunogold particles (around 1 nm diameter) should be used to facilitate penetration of the particle into the tissue and access to subcellular domains. As such small particles cannot be visualized in the light microscope, a silver enhancement step is required to increase the diameter of the particles. Several silver enhancement kits are commercially available. The duration of the enhancement step can be adjusted to yield particles of differing sizes. Although it is possible to use differential silver enhancement to perform multiple immunogold staining, this is usually reserved for electron microscopy studies, as the different sizes can be difficult to resolve in the light microscope. Instead, immunogold particle staining may be combined with immunoenzyme staining at the light microscope level (see *Fig. 7B*, also available in the color section).

2.2 Labels for fluorescence and confocal microscopy techniques

As mentioned in Section 1.2.2, there is a wide range of fluorescent labels available for use in single and multiple staining studies. The choice of probes will depend primarily on the microscope system to be used. This will define which fluorochromes can be excited and which emission wavelengths can be detected. Most fluorescent/confocal microscopes are compatible with fluorescein isothiocyanate (FITC)- and rhodamine-like fluorochromes, making this a common combination in double staining experiments (see *Fig. 7C*, also available in the color section). The added use of UV- and far red light-excitable fluorochromes enables triple or even quadruple staining (9–11). Interestingly, it has been reported that immunoenzyme double staining and immunofluorescence double staining may also be combined within the same sample to yield quadruple staining of target proteins (12).

2.3 Labels for electron microscopy techniques

Immunoelectron microscopy is not a common technique; however, it can be a valuable tool for analyzing the subcellular distribution of proteins within tissue samples (see Chapter 7, section 1.4). Many labels used in light microcopy, such as the immunoperoxidase substrate DAB and immunogold particles, are compatible with electron microscopy, as they are electron dense and thus readily visible within the electron microscope. It is difficult to distinguish multiple immunoenzyme substrates within the electron microscope, as they can only be identified by physical appearance and not by color, as is possible in the light microscope. Therefore, multiple immunoenzyme staining is rare at the electron microscope level. Instead, immunoenzyme + immunogold (see *Fig. 7F*, also available in the color section) or multiple immunogold (see *Fig. 7D* and *E*, also available in the color section) staining may be performed. Both of these methods are compatible with pre-embedding electron microscopy, where the tissue is immunochemically stained prior to embedding in resin and cutting ultrathin sections (9, 13). Post-embedding electron microscopy, in which immunochemical staining is performed on resin-embedded ultrathin sections, is more suited to multiple immunogold staining, as the resin etch step is deleterious to immunoenzyme substrates, leaching them from the section (14). The great advantage of post-embedding staining is that, as the sections are surface labeled, a wide range of immunogold particle sizes can be used, facilitating the co-localization of several proteins on the same section. This is harder to achieve with pre-embedding staining where differential silver enhancement is used to differentiate individual immunoreactivities, resulting in a range of particles sizes.

2.4 Recommended protocols

This section provides basic, robust protocols for multiple immunochemical staining at the light, fluorescence (confocal), and electron microscope levels. The methods described should be considered as starting points and may need optimizing for particular target proteins. Issues that may need to be addressed include:

- **Antigen retrieval.** Under certain fixation conditions, the epitopes on many protein targets are masked, hindering binding of the primary antibody. There are a variety of enzymatic and heat-induced antigen retrieval steps that can be employed at the start of the protocol to facilitate antibody binding (15) (see Chapter 4, section 2.10.1). If it is necessary to utilize either of these procedures to visualize one of the target proteins, it is important to verify, using single staining, that immunoreactivity for the other target proteins is unaltered by the retrieval method.
- **Permeablization.** Most immunochemical applications require a permeablization step to facilitate immunoreagent access into the sample (see Chapter 4, section 2.1.4, and section 2.10.2, *Protocol 4*, note b).

- **Antibody incubation times.** Antibody incubation times are dependant on the thickness and nature of the tissue sample. As a rough guide, allow 1 h per 10 mm sample thickness. Overnight incubations can be performed at 4°C to reduce microbial growth and contamination (see Chapter 4, section 2.10.2, *Protocol 4*, note g).

Protocol 1

Double immunoenzyme staining (light microscopy)

Equipment and Reagents

Please refer to *Appendix 1* for recipes.

- 50 mM TBS (pH 7.4)
- 50 mM TBS (pH 8.0) containing 0.05% (w/v) DAB[a] (also see Chapter 3, section 2.1.1) + 0.01% (v/v) H_2O_2
- 50 mM TBS (pH 8.0) containing 0.05% (w/v) DAB[a] + 0.375% (w/v) nickel sulphate + 0.01% (v/v) H_2O_2
- 100% Industrial methylated spirits (IMS) or methanol
- Appropriate staining apparatus
- ABC–HRP complex (see Chapter 3, section 2.1.1, and Chapter 4, *Protocol 4*, note i) in PBS (made up following the manufacturer's instructions)
- Coverslips (see Chapter 3, section 2.5)
- Humidified incubation chamber (see general guidelines discussed in Chapter 4, section 2.10.2)
- Mounting media (see Chapter 4, section 2.7)
- Normal serum from the species used to raise the secondary antibody
- Phosphate-buffered saline (PBS)
- PBS containing 0.1% (v/v) Triton X-100 (see Chapter 4, *Protocol 4*, note c)
- PBS containing 1% (w/v) sodium borohydride[b]
- PBS containing 10% (v/v) normal serum + 0.5% (w/v) BSA (see Chapter 4, *Protocol 4*, note d)
- PBS containing 0.06% (v/v) H_2O_2[c] (also see Chapter 4, *Protocol 4*, note j)
- Primary antibodies optimally diluted in PBS containing 1% (v/v) normal serum + 0.5% (w/v) bovine serum albumin (BSA)
- Secondary antibodies optimally diluted in PBS containing 1% (v/v) normal serum + 0.5% (w/v) BSA
- Xylene (or other dewaxing/clearing reagent) (see Chapter 4, *Protocol 4*, note n)

Method

All steps are carried out at room temperature

1. Immerse the slides for 1 min in PBS.
2. Immerse the slides for 20 min in PBS containing 0.1% (v/v) Triton X-100.
3. Immerse the slides in PBS containing 0.06% (v/v) H_2O_2[c] for 30 min.
4. Immerse the slides three times for 10 min each in PBS.
5. Immerse the slides in PBS containing 1% (w/v) sodium borohydride[b] for 20 min.

6. Immerse the slides three times for 10 min each in PBS.
7. Incubate the slides in PBS containing 10% (v/v) normal serum + 0.5% (w/v) BSA for 30 min.
8. Immerse the slides for 1 min in PBS.
9. For the primary antibody co-incubation, incubate the slides with both primary antibodies optimally diluted in PBS containing 1% (v/v) normal serum + 0.5% (w/v) BSA for 1 h to overnight.
10. Immerse the slides three times for 10 min each in PBS.
11. For the first secondary antibody incubation, incubate the slides with biotinylated antibody raised against the species of the first primary antibody, optimally diluted in PBS containing 1% (v/v) normal serum + 0.5% (w/v) BSA for 1 h to overnight.
12. Immerse the slides three times for 10 min each in PBS.
13. Incubate the slides in ABC–HRP complex (or equivalent) in PBS (made up following the manufacturer's instructions) for 1 h.
14. Immerse the slides twice for 10 min each in PBS.
15. Immerse the slides for 10 min in 50 mM TBS (pH 7.4).
16. Incubate the slides in 50 mM TBS (pH 8.0) containing 0.05% (w/v) DAB[a] + 0.01% (v/v) H_2O_2. Monitor color development carefully (typically for 2 to 10 min).
17. Immerse the slides three times for 10 min each in 50 mM TBS (pH 7.4).
18. Incubate slides in PBS containing 0.06% (v/v) H_2O_2[c] for 30 min.
19. Immerse the slides three times for 10 min each in PBS.
20. Repeat steps 10–13 using biotinylated secondary antibody raised against the species of the second primary antibody.
21. Incubate the slides in 50 mM TBS (pH 8.0) containing 0.05% (w/v) DAB[a] + 0.375% (w/v) nickel sulphate + 0.01% (v/v) H_2O_2. Monitor color development carefully (typically for 2 to 10 min).
22. To terminate the reaction, immerse the slides three times for 10 min each in TBS (pH 7.4).
23. Samples can either be mounted in an aqueous mounting medium or dehydrated through an increasing alcohol series, cleared, and mounted in an organic mounting medium such as DPX.
24. View using light microscopy.

Notes

[a]Caution must be taken whilst working with DAB as it is a suspected carcinogen. Dispose of any waste DAB appropriately (consult local guidelines).

[b]Sodium borohydride lowers background staining by blocking free aldehyde groups, reducing –CHO groups to –OH. It can be damaging to tissues due to the release of hydrogen gas, so monitor the reaction. It is not always required in paraformaldehyde-fixed tissue but is often necessary when glutaraldehyde fixation is used.

[c]Hydrogen peroxide may be used to quench any endogenous peroxidase activity in the sample, as is found, for instance, in erythrocytes. If endogenous enzyme activity is not a problem, this step may be omitted.

Protocol 2
Multiple immunofluorescence staining (fluorescence microscopy)

Equipment and Reagents

Please refer to *Appendix 1* for recipes.

- Antifadant-containing aqueous mounting media
- Appropriate staining apparatus
- Clear nail varnish
- Coverslips (see Chapter 3, section 2.5)
- Humidified incubation chamber (see general guidelines discussed in Chapter 4, section 2.10.2)
- Normal serum from the species used to raise the secondary antibody
- PBS
- PBS containing 0.1% (v/v) Triton X-100 (see Chapter 4, *Protocol 4*, note c)
- PBS containing 0.3 M glycine
- PBS containing 10% (v/v) normal serum + 0.5% (w/v) BSA (see Chapter 4, *Protocol 4*, note d)
- Primary antibodies optimally diluted in PBS containing 1% (v/v) normal serum + 0.5% (w/v) BSA
- Secondary fluorochrome-conjugated antibodies optimally diluted in PBS containing 1% (v/v) normal serum + 0.5% (w/v) BSA
- Xylene (or other dewaxing/clearing reagent) (see Chapter 4, *Protocol 4*, note n)

Method

1. Immerse the slides for 1 min in PBS.
2. Immerse the slides for 20 min in PBS containing 0.1% (v/v) Triton X-100.
3. Immerse the slides three times for 10 min each in PBS.
4. Incubate in PBS containing 0.3 M glycine[a] for 20 min.
5. Immerse the slides three times for 10 min each in PBS.
6. Incubate the slides in PBS containing 10% (v/v) normal serum + 0.5% (w/v) BSA for 30 min.
7. Immerse the slides for 1 min in PBS.
8. For the primary antibody co-incubation, incubate the slides with both primary antibodies optimally diluted in PBS containing 1% normal serum (v/v) + 0.5% (w/v) BSA for 1 h to overnight.
9. Immerse the slides three times for 10 min each in PBS.
10. For the secondary fluorochrome-conjugated antibody co-incubation[b], incubate the slides with both secondary antibodies optimally diluted in PBS containing 1% (v/v) normal serum + 0.5% (w/v) BSA for 1 h to overnight in the dark.
11. Immerse the slides three times for 10 min each in PBS in the dark.
12. Mount in antifadant-containing mounting media, coverslip, and seal with nail varnish.
13. View using fluorescence or confocal microscopy.

Notes

[a]Glycine reduces background fluorescence by binding to free aldehyde groups, which could otherwise bind covalently with side-chain lysine groups on antibodies or enzyme labels.
[b]The choice of fluorochrome-conjugated secondary antibodies will depend on the microscope system to be used. Before starting, check that the system is capable of exciting each fluorochrome and detecting the individual emission spectra.

Protocol 3

Double immunogold pre-embedding staining (electron microscopy)

Equipment and Reagents

Please refer to *Appendix 1* for recipes.

- 20 mM Sodium citrate buffer (pH 7.0)
- 30 mM Aqueous sodium thiosulfate
- Appropriate staining apparatus
- Distilled water
- Humidified incubation chamber (see general guidelines discussed in Chapter 4, section 2.10.2)
- Normal serum from the species used to raise the secondary antibody
- Phosphate buffer (PB): 0.1 M sodium phosphate (pH 7.4)
- PB containing 0.1% (w/v) sodium borohydride[a]
- PB containing 0.5% (w/v) osmium tetroxide[b]
- PB containing 2.5% (v/v) glutaraldehyde[b]
- PBS containing 0.05% (v/v) Triton X-100 (see Chapter 4, *Protocol 4*, note c)
- PBS containing 0.2% (w/v) BSA
- PBS containing 10% normal serum + 0.5% (w/v) BSA (see Chapter 4, *Protocol 4*, note d)
- Primary antibodies optimally diluted in PBS containing 0.2% (w/v) BSA
- Resin for embedding
- Secondary 1 nm gold-conjugated antibodies optimally diluted in PBS containing 0.2% (w/v) BSA
- Silver enhancement reagent (refer to manufacturer's datasheet)

Method

1. Incubate slides in PB containing 0.1% (w/v) sodium borohydride[a] for 15 min.
2. Immerse the slides four times for 10 min each in PB.
3. Immerse the slides for 30 min in PBS containing 0.05% (v/v) Triton X-100.
4. Immerse the slides four times for 10 min each in PB.
5. Incubate the slides in PBS containing 10% (v/v) normal serum + 0.5% (w/v) BSA for 30 min.
6. Immerse the slides twice for 10 min each in PBS containing 0.2% (w/v) BSA.
7. For primary antibody co-incubation, incubate the slides with both primary antibodies optimally diluted in PBS containing 0.2% (w/v) BSA overnight at 4°C.
8. Immerse the slides six times for 10 min each in PBS containing 0.2% (w/v) BSA.

9. For the first secondary antibody incubation, incubate the slides with the 1 nm gold-conjugated antibody raised against the species of the first primary antibody, optimally diluted in PBS containing 0.2% (w/v) BSA, overnight at 4°C.
10. Immerse the slides six times for 10 min each in PBS containing 0.2% (w/v) BSA.
11. Immerse the slides twice for 10 min each in PB.
12. Immerse the slides twice for 10 min each in 20 mM sodium citrate buffer (pH 7.0).
13. For the first silver enhancement[d], incubate the slides with the first silver enhancement solution to enhance immunogold particles to approximately 10 nm diameter (refer to manufacturer's datasheet for recommended enhancement time).
14. For enhancement termination, immerse the slides for 10 min in 30 mM aqueous sodium thiosulfate.
15. Immerse the slides twice for 10 min each in distilled water.
16. Immerse the slides twice for 10 min each in PBS.
17. For the second secondary antibody incubation, incubate the slides with the 1 nm gold conjugate directed against the species of second primary antibody, diluted accordingly in 0.2% (w/v) BSA in PBS, overnight at 4°C.
18. Immerse the slides six times for 10 min each in PBS containing 0.2% (w/v) BSA.
19. Immerse the slides twice for 10 min each in PB.
20. For post-fixation, incubate the slides in PB containing 2.5% (v/v) glutaraldehyde[b] for 2 h.
21. Immerse the slides twice for 10 min each in PB.
22. Immerse the slides four times for 10 min each in distilled water.
23. For the second silver enhancement[d], incubate the slides with the second silver enhancement solution to enhance the immunogold particles to approximately 10 nm diameter (refer to manufacturer's datasheet for recommended enhancement time).
24. For enhancement termination, immerse the slides for 10 min in 30 mM aqueous sodium thiosulfate.
25. Immerse the slides four times for 10 min each in distilled water.
26. For osmication, incubate in PB containing 0.5% (w/v) osmium tetroxide[b] for 15 min.
27. The tissues are now ready for resin embedding[c] and processing for electron microscopy (see Chapter 7, section 2).

Notes

[a]Sodium borohydride lowers background staining by reducing free aldehyde groups. It can be damaging to tissues due to the release of hydrogen gas, so monitor the reaction. It is not always required in PFA-fixed tissue but often necessary when glutaraldehyde fixation is used.

[b]Glutaraldehyde and osmium tetroxide must be used in a fume hood. Post-fixation in glutaraldehyde helps to enhance tissue morphology. Osmication performs a similar role, but with the advantage of adding positive contrast to the tissue, as it is a heavy metal that scatters electrons (see Chapter 7, section 2.1).

[c]Any one of a variety of electron microscopy resins may be used. It is recommended to try an epoxy-based resin in the first instance.

[d]Silver enhancement is a form of signal amplification for gold labels. Silver is reduced in the presence of gold, leading to a build-up of silver on the surface of the gold labels, increasing their size and visibility, typically amplifying the signal by between 10 and 100 times. The enhancement reaction is terminated by immersing the slides in 30 mM aqueous sodium thiosulfate.

Protocol 4

Immunogold + immunoenzyme double pre-embedding staining (light and electron microscopy)

Equipment and Reagents

Please refer to *Appendix 1* for recipes.

- 30 mM Aqueous sodium thiosulfate
- 50 mM Tris buffer (pH 7.4)
- ABC-HRP complex (see Chapter 3, section 2.2.1, and Chapter 4, *Protocol 4*, note i) in PBS (made up following the manufacturer's instructions)
- Appropriate staining apparatus
- Biotinylated secondary antibody and 1 nm gold-conjugated secondary antibody against the first primary antibody species, both optimally diluted in PBS containing 0.2% (w/v) BSA
- Tris buffer containing 0.05% (w/v) DAB[a] + 0.01% (v/v) H_2O_2
- Distilled water
- For electron microscopy: PB containing 0.5% (w/v) osmium tetroxide[b] and resin for embedding
- For light microscopy: coverslips (see Chapter 3, section 2.5)
- For light microscopy: mounting media (see Chapter 4, section 2.7)
- Humidified incubation chamber (see general guidelines discussed in Chapter 4, section 2.10.2)
- Normal serum from the species used to raise the secondary antibody
- Phosphate buffer (PB): 0.1 M sodium phosphate (pH 7.4)
- PB containing 0.1% (w/v) sodium borohydride[c]
- PB containing 1% (v/v) glutaraldehyde[b]
- PBS
- PBS containing 0.05% (v/v) Triton X-100 (see Chapter 4, *Protocol 4*, note c)
- PBS containing 0.2% (w/v) BSA
- PBS containing 10% (v/v) normal serum + 0.2% (w/v) BSA (see Chapter 4, *Protocol 4*, note d)
- Primary antibodies optimally diluted in PBS containing 0.2% (w/v) BSA
- Silver enhancement reagent (refer to manufacturer's datasheet)
- Xylene (or other dewaxing/clearing reagent) (see Chapter 4, *Protocol 4*, note n)

Method

1. Incubate the slides in PB containing 0.1% (w/v) sodium borohydride[c] for 15 min.
2. Immerse the slides four times for 10 min each in PB.
3. Immerse the slides for 30 min in PBS containing 0.05% (v/v) Triton X-100.
4. Immerse the slides four times for 10 min each in PB.
5. Incubate the slides in PBS containing 10% (v/v) normal serum + 0.2% (w/v) BSA for 30 min.
6. Immerse the slides twice for 10 min each in PBS containing 0.2% (w/v) BSA.
7. For primary antibody co-incubation, incubate the slides with both primary antibodies, optimally diluted in PBS containing 0.2% (w/v) BSA, overnight at 4°C.
8. Immerse the slides six times for 10 min each in PBS containing 0.2% (w/v) BSA.
9. For multiple secondary antibody incubations, incubate the slides with biotinylated antibody directed against the species of the first primary antibody and 1 nm gold

conjugate directed against the species of the second primary antibody, both diluted accordingly in 0.2% (w/v) BSA in PBS, overnight at 4°C.

10. Immerse the slides four times for 10 min each in PBS containing 0.2% (w/v) BSA.
11. Immerse the slides for 10 min in PBS.
12. Immerse the slides for 10 min in PB.
13. For post-fixation to immunobilize the gold particles, incubate the slides in PB containing 1% (v/v) glutaraldehyde[b] for 10 min.
14. Immerse the slides two times for 10 min each in PB.
15. Immerse the slides four times for 10 min each in distilled water.
16. For silver enhancement[d], enhance the immunogold particles to approximately 10 nm diameter (refer to the manufacturer's datasheet for the recommended enhancement time).
17. For enhancement termination, immerse the slides for 10 min in 30 mM aqueous sodium thiosulfate.
18. Immerse the slides four times for 10 min each in distilled water.
19. Immerse the slides twice for 10 min each in PBS.
20. Incubate the slides in ABC HRP complex in PBS (made up following the manufacturer's instructions) for 2 h.
21. Immerse the slides twice for 10 min each in PBS.
22. Immerse the slides for 10 min in 50 mM Tris buffer (pH 7.4).
23. Incubate the slides in Tris buffer containing 0.05% (w/v) DAB[a] + 0.01% (v/v) H_2O_2. Monitor color development carefully (typically for 2 to 10 min).
24. Immerse slides for three times 10 min in Tris buffer (pH 7.4).
25. For sections for light microscopy analysis, if necessary, transfer samples to clean microscope slides and air dry in a dust-free environment. Samples can be mounted in an aqueous mounting medium or dehydrated through an increasing alcohol series, cleared in solvent, and mounted in a solvent-based medium such as DPX.
26. For osmication of sections for electron microscopy, incubate for 15 min in PB containing 0.5% (w/v) osmium tetroxide[b], followed by embedding in the preferred resin (see Chapter 7, section 2)

Notes

[a]Caution must be taken whilst working with DAB as it is a suspected carcinogen. Dispose of any waste DAB appropriately (consult local guidelines).

[b]Glutaraldehyde and osmium tetroxide must be used in a fume hood. Post-fixation in glutaraldehyde helps to enhance tissue morphology. Osmication performs a similar role, but with the advantage of adding positive contrast to the tissue, as it is a heavy metal that scatters electrons (see Chapter 7, section 2.1)

[c]Sodium borohydride lowers background staining by reducing free aldehyde groups. It can be damaging to tissues due to the release of hydrogen gas, so monitor the reaction. It is not always required in paraformaldehyde-fixed tissue but is often necessary when glutaraldehyde fixation is used.

[d]Silver enhancement is a form of signal amplification for gold labels. Silver is reduced in the presence of gold, leading to a build-up of silver on the surface of the gold labels, increasing their size and visibility, typically amplifying the signal by between 10 and 100 times. The enhancement reaction is terminated by immersing the slides in 30 mM aqueous sodium thiosulfate.

Protocol 5

Multiple immunogold post-embedding staining using acrylate-based resins (electron microscopy)

Equipment and Reagents

Please refer to *Appendix 1* for recipes.

- Appropriate staining apparatus
- Nickel grids
- Distilled water
- Distilled water containing 1% (w/v) uranyl acetate
- Humidified incubation chamber (see general guidelines discussed in Chapter 4, section 2.10.2)
- Normal human serum (NHS)
- Phosphate buffer (PB): 0.1 M sodium phosphate (pH 7.4)
- PB containing 2% (v/v) glutaraldehyde[a]
- Primary antibodies optimally diluted in TBS containing 2% (v/v) NHS + 0.01% (v/v) Triton X-100
- Reynolds lead citrate
- Secondary antibodies optimally diluted in TBS containing 2% (v/v) NHS + 0.01% (v/v) Triton X-100 + 0.05% (v/v) polyethylene glycol
- Sodium ethanolate[b]
- TBS (pH 7.4)
- TBS containing 2% (v/v) NHS[c] + 0.01% (v/v) Triton X-100

Method

1. For sample preparation (see Chapter 7, section 2), collect ultrathin sections on nickel grids.
2. For resin etching, incubate the grids for 3 s in saturated sodium ethanolate[b] solution.
3. Dip the grids in three changes of distilled water and place them on individual drops of distilled water.
4. Immerse the grids twice for 5 min each in TBS, pH 7.4.
5. Incubate the grids for 30 min in TBS containing 2% (v/v) NHS[c] + 0.01% (v/v) Triton X-100.
6. For primary antibody co-incubation, incubate the grids with both primary antibodies optimally diluted in TBS containing 2% (v/v) NHS + 0.01% (v/v) Triton X-10, overnight at 4°C.
7. Immerse the grids three times for 10 min each in TBS (pH 7.4).
8. For the secondary gold-conjugated antibody co-incubation, incubate the grids with secondary antibodies optimally diluted in TBS containing 2% (v/v) NHS + 0.01% (v/v) Triton X-100 + 0.05% (v/v) polyethylene glycol for 2 h.
9. Immerse the grids three times for 10 min each in TBS (pH 7.4).
10. Immerse the grids once for 10 min in PB.
11. For post-fixation to immunobilize the gold particles, incubate the grids in PB containing 2% (v/v) glutaraldehyde[a] for 2 min.

12. Dip the grids in three changes of distilled water and place the grids on individual drops of water.
13. Air dry.
14. To counterstain, incubate the grids in distilled water containing 1% (w/v) uranyl acetate for 35 min in the dark.
15. Dip the grids in three changes of distilled water and place the grids on individual drops of water.
16. Air dry.
17. Incubate the grids in Reynolds lead citrate[d] for 3 min.
18. Dip the grids in five changes of distilled water and air dry.
19. View using an electron microscope.

Notes

[a]Glutaraldehyde must be used in a fume hood.
[b]Sodium ethanolate should be prepared at least 24 h in advance by adding an excess of sodium ethanolate to absolute ethanol in a glass bottle. Store in the dark in a solvent or fume cupboard.
[c]Normal human serum has been found to be an effective, albeit costly, blocking reagent in post-embedding labeling. Normal serum from the same species as the secondary antibodies may be supplemented if desired.
[d]Reynolds lead citrate is used as a counterstain in electron microscopy (16).

Protocol 6

Multiple immunofluorescence labeling using same-species primary antibodies: the use of conjugated Fab fragments for blocking and labeling (see *Fig. 4*)

Equipment and Reagents

Please refer to *Appendix 1* for recipes.

- Antifadant mounting media (see Chapter 4, section 2.7)
- Appropriate labeling apparatus
- Clear nail varnish
- Coverslips (see Chapter 3, section 2.5)
- First primary antibody optimally diluted in PBS containing 1% (v/v) normal serum + 0.5% (w/v) BSA
- First secondary Fab fragment, conjugated to fluorochrome #1, optimally diluted in PBS containing 0.5% (w/v) BSA + 1% (v/v) normal serum
- Humidified incubation chamber (see general guidelines discussed in Chapter 4, section 2.10.2)
- Normal serum from the same species as the secondary Fab fragments
- PBS

- PBS containing 0.1% (v/v) Triton X-100 (see Chapter 4, *Protocol 4*, note c)
- PBS containing 0.3 M glycine
- PBS containing 10% (v/v) normal serum + 0.5% (w/v) BSA (see Chapter 4, *Protocol 4*, note d)
- Second primary antibody optimally diluted in PBS containing 1% (v/v) normal serum + 0.5% (w/v) BSA
- Second secondary Fab fragment, conjugated to fluorochrome #2, optimally diluted in PBS containing 0.5% (w/v) BSA + 1% (v/v) normal serum

Method

1. Immerse the slides once for 1 min in PBS.
2. Immerse the slides in PBS containing 0.1% Triton X-100 for 20 min.
3. Immerse the slides three times for 10 min each in PBS.
4. Incubate the slides in PBS containing 0.3 M glycine[a] for 20 min.
5. Immerse the slides three times for 10 min each in PBS.
6. Incubate slides in PBS containing 10% (v/v) normal serum + 0.5% (w/v) BSA for 30 min.
7. Immerse the slides once for 1 min in PBS.
8. For the first primary antibody incubation, incubate the slides with primary antibody optimally diluted in PBS containing 1% (v/v) normal serum + 0.5% (w/v) BSA for 1 h to overnight.
9. Immerse the slides three times for 10 min each in PBS.
10. For the first secondary Fab fragment[b] incubation, incubate the slides with fluorochrome #1-conjugated Fab fragment raised against the species of the first primary antibody, optimally diluted in PBS containing 0.5% (w/v) BSA + 1% (v/v) normal serum, overnight at 4°C.
11. Immerse the slides three times for 10 min each in PBS.
12. For the second primary antibody incubation, incubate the slides with primary antibody optimally diluted in PBS containing 1% (v/v) normal serum + 0.5% (w/v) BSA for 1 h to overnight.
13. Immerse the slides three times for 10 min each in PBS.
14. For the second secondary Fab fragment[b] incubation, incubate the slides with fluorochrome #2-conjugated Fab fragment raised against the species of the second primary antibody, optimally diluted in PBS containing 0.5% (w/v) BSA + 1% (v/v) normal serum, overnight at 4°C.
15. Immerse the slides three times for 10 min each in PBS.
16. Mount in antifadant mounting media, coverslip, and seal with nail varnish.
17. View using fluorescence or confocal microscopy.

Notes

[a]Glycine reduces background fluorescence by binding to free aldehyde groups.
[b]The choice of fluorochrome-conjugated secondary reagents will depend on the microscope system to be used. Before starting, check that the system is capable of exciting each fluorochrome and detecting the individual emission spectra.

Protocol 7

Multiple immunofluorescence labeling using same-species primary antibodies: the use of unconjugated Fab fragments to convert the first primary antibody into a different species (see *Fig. 5*)

Equipment and Reagents

Please refer to *Appendix 1* for recipes.

- Antifadant mounting media (see Chapter 4, section 2.7)
- Appropriate labeling apparatus
- Clear nail varnish
- Coverslips (see Chapter 3, section 2.5)
- First primary antibody optimally diluted in PBS containing 1% (v/v) normal serum + 0.5% (w/v) BSA
- Fluorochrome #1-conjugated secondary antibody raised against the species of the first Fab fragment, optimally diluted in PBS containing 0.5% (w/v) BSA + 1% (v/v) normal serum
- Fluorochrome #2-conjugated secondary antibody raised against the species of the second primary antibody, optimally diluted in PBS containing 0.5% (w/v) BSA + 1% (v/v) normal serum
- Humidified incubation chamber (see general guidelines discussed in Chapter 4, section 2.10.2)
- Normal serum from the species used to raise the secondary antibodies
- PBS
- PBS containing 0.1% (v/v) Triton X-100 (see Chapter 4, *Protocol 4*, note c)
- PBS containing 0.3 M glycine[a]
- PBS containing 10% (v/v) normal serum + 0.5% (w/v) BSA (see Chapter 4, *Protocol 4*, note d)
- Second primary antibody optimally diluted in PBS containing 1% (v/v) normal serum + 0.5% (w/v) BSA
- Unconjugated Fab fragment raised against the species of the first primary antibody, optimally diluted in PBS containing 1% (v/v) normal serum + 0.5% (w/v) BSA

Method

1. Immerse the slides for 1 min in PBS.
2. Immerse the slides for 20 min in PBS containing 0.1% (v/v) Triton X-100.
3. Immerse the slides three times for 10 min each in PBS.
4. Incubate the slides in PBS containing 0.3 M glycine[a] for 20 min.
5. Immerse the slides three times for 10 min each in PBS.
6. Incubate slides in PBS containing 10% (v/v) normal serum + 0.5% (w/v) BSA for 30 min.
7. Immerse the slides for 1 min in PBS.
8. For the first primary antibody incubation, incubate the slides with primary antibody optimally diluted in PBS containing 1% (v/v) normal serum + 0.5% (w/v) BSA for 1 h to overnight.

9. Immerse the slides three times for 10 min each in PBS.
10. For the unconjugated Fab fragment incubation, incubate the slides with unconjugated Fab fragment raised against the species of the first primary antibody, optimally diluted in PBS containing 1% (v/v) normal serum + 0.5% (w/v) BSA for 2 h.
11. Immerse the slides three times for 10 min each in PBS.
12. For the fluorochrome #1-conjugated secondary antibody incubation[b], incubate the slides with fluorochrome #1-conjugated secondary antibody raised against the species of the first Fab fragment, optimally diluted in PBS containing 0.5% (w/v) BSA + 1% (v/v) normal serum for 1 h to overnight.
13. Immerse the slides three times for 10 min each in PBS.
14. For the second primary antibody incubation, incubate the slides with primary antibody optimally diluted in PBS containing 1% (v/v) normal serum + 0.5% (w/v) BSA for 1 h to overnight.
15. Immerse the slides three times for 10 min each in PBS.
16. For the fluorochrome #2-conjugated secondary antibody incubation[b], incubate the slides with fluorochrome #2-conjugated secondary antibody raised against the species of the second primary antibody, optimally diluted in PBS containing 0.5% (w/v) BSA + 1% (v/v) normal serum for 1 h to overnight.
17. Immerse the slides three times for 10 min each in PBS.
18. Mount in antifadant mounting media, coverslip, and seal with nail varnish.
19. View using fluorescence or confocal microscopy.

Notes

[a]Glycine reduces background fluorescence by binding to free aldehyde groups.
[b]The choice of fluorochrome-conjugated secondary reagents will depend on the microscope system to be used. Before starting, check that the system is capable of exciting each fluorochrome and detecting the individual emission spectra.

Protocol 8

Multiple immunofluorescence labeling using same-species primary antibodies: the use of unconjugated Fab fragments to block after the first secondary antibody step (see *Fig. 6*)

Equipment and Reagents
Please refer to *Appendix 1* for recipes.

- Antifadant mounting media (see Chapter 4, section 2.7)
- Appropriate labeling apparatus
- Clear nail varnish
- Coverslips (see Chapter 3, section 2.5)

- Fluorochrome #1-conjugated secondary antibody raised against the species of the first primary antibody, optimally diluted in PBS containing 0.5% (w/v) BSA + 1% (v/v) normal serum (secondary antibody species)
- Fluorochrome #2-conjugated secondary antibody raised against the species of the second primary antibody, optimally diluted in PBS containing 0.5% (w/v) BSA + 1% (v/v) normal serum (secondary antibody species)
- Humidified incubation chamber (see general guidelines discussed in Chapter 4, section 2.10.2)
- Normal serum from the species used to raise both the primary and secondary antibodies
- PBS
- PBS containing 0.1% (v/v) Triton X-100 (see Chapter 4, *Protocol 4*, note d)
- PBS containing 0.3 M glycine[a]
- PBS containing 10% (v/v) normal serum (primary antibody species)
- PBS containing 10% (v/v) normal serum (secondary antibody species) + 0.5% (w/v) BSA
- Primary antibody optimally diluted in PBS containing 1% (v/v) normal serum (secondary antibody species) + 0.5% (w/v) BSA
- For the second primary antibody incubation: primary antibody optimally diluted in PBS containing 1% (v/v) normal serum (secondary antibody species) + 0.5% (w/v) BSA
- Unconjugated Fab fragment raised against the species of the first primary antibody, optimally diluted in PBS containing 0.5% (w/v) BSA + 1% (v/v) normal serum (secondary antibody species)

Method

1. Immerse the slides for 1 min in PBS.
2. Immerse the slides for 20 min in PBS containing 0.1% (v/v) Triton X-100.
3. Immerse the slides three times for 10 min each in PBS.
4. Incubate the slides in PBS containing 0.3 M glycine[a] for 20 min.
5. Immerse the slides three times for 10 min each in PBS.
6. Incubate the slides in PBS containing 10% (v/v) normal serum (secondary antibody species) + 0.5% (w/v) BSA for 30 min.
7. Immerse the slides for 1 min in PBS.
8. For the first primary antibody incubation, incubate the slides with primary antibody optimally diluted in PBS containing 1% (v/v) normal serum (secondary antibody species) + 0.5% (w/v) BSA for 1 h to overnight.
9. Immerse the slides three times for 10 min each in PBS.
10. For the fluorochrome #1-conjugated secondary antibody incubation[b], incubate the slides with fluorochrome #1-conjugated secondary antibody raised against the species of the first primary antibody, optimally diluted in PBS containing 0.5% (w/v) BSA + 1% (v/v) normal serum (secondary antibody species) for 1 h to overnight.
11. Immerse the slides three times for 10 min each in PBS.
12. Incubate the slides in PBS containing 10% (v/v) normal serum (primary antibody species) for 30 min.

13. Immerse the slides three times for 10 min each in PBS.
14. For the unconjugated Fab fragment incubation, incubate the slides with unconjugated Fab fragment raised against the species of the first primary antibody, optimally diluted in PBS containing 0.5% (w/v) BSA + 1% (v/v) normal serum (secondary antibody species) for 2 h.
15. Immerse the slides three times for 10 min each in PBS.
16. For the second primary antibody incubation, incubate the slides with primary antibody optimally diluted in PBS containing 1% (v/v) normal serum (secondary antibody species) + 0.5% (w/v) BSA for 1 h to overnight.
17. Immerse the slides three times for 10 min each in PBS.
18. For the fluorochrome #2-conjugated secondary antibody incubation[b], incubate the slides with fluorochrome #2-conjugated secondary antibody raised against the species of the second primary antibody, optimally diluted in PBS containing 0.5% (w/v) BSA + 1% (v/v) normal serum (secondary antibody species) for 1 h to overnight.
19. Immerse the slides three times for 10 min each in PBS.
20. Mount in antifadant mounting media, coverslip, and seal with nail varnish.
21. View using fluorescence or confocal microscopy.

Notes

[a]Glycine reduces background fluorescence by binding to free aldehyde groups.
[b]The choice of fluorochrome-conjugated secondary reagents will depend on the microscope system to be used. Before starting, check that the system is capable of exciting each fluorochrome and detecting the individual emission spectra.

3. TROUBLESHOOTING (for a full troubleshooting guide, see Chapter 9)

The protocols described in this chapter provide robust starting points for undertaking a range of multiple immunochemical staining experiments at the light, fluorescence, and electron microscope levels. As with any immunochemical staining strategy, however, the methods are not set in stone and are amenable to modifications to suit individual samples, fixation regimes, and detection system combinations. The most critical element is an appreciation and understanding of the individual steps contributing to the staining process (see *Fig. 2*). By taking the time to choose an appropriate protocol and through careful design of the staining stages, many of the potential pitfalls often encountered in multiple staining experiments, such as those listed below, can be avoided.

No staining or weak signal

- Increase the primary and/or secondary antibody concentrations, checking for improved immunoreactivity in single-labeled controls.
- Incorporate an antigen retrieval step, such as heat-induced epitope retrieval or protease-induced antigen retrieval (15), prior to the blocking step.
- Try an alternative fixation regime.
- Modify the permeabilization steps, either by adding Triton X-100 to the antibody incubation medium or by using an alternative permeabilization agent such as saponin (17).
- Check the pH of all buffers (should be around pH 7.4).

High background staining

- Decrease the primary and/or secondary antibody concentrations, checking for improved immunoreactivity in single-labeled controls.
- Centrifuge all immunoreagent solutions prior to use to remove any aggregates, which may stick to the samples.
- Increase the duration of nonspecific-binding blocking steps.
- Incorporate additional blocking steps to reduce endogenous sources of background staining:
  - For peroxidase-based labels, add an additional endogenous peroxidase block, such as 0.3% H_2O_2 in buffer, prior to antibody incubations.
  - For AP-based labels, add 10 mM levamisol to the substrate solution.
  - For avidin–biotin-based labels, add an endogenous biotin blocking step prior to antibody incubations (kits are available from most immunochemical suppliers).
  - If the primary antibody species is the same as the sample, for instance using a mouse antibody on mouse sections, employ an endogenous immunoglobulin-blocking step (kits are available from most immunochemical suppliers).

Inability to distinguish individual immunoreactivities in multiple-labeled samples

- Employ a different multiple staining protocol. Often it is difficult to resolve individual immunoreactivities if the target proteins are within the same cellular compartment, resulting in signal overlap. This is a particular problem in multiple immunoenzyme techniques where the presence of a dark substrate, such as nickel–DAB, can mask a lighter substrate, such as DAB. In this instance, multiple immunofluorescence may be a more suitable alternative.

Unexpected overlap of individual immunoreactivities

- Compare immunoreactivities with single-labeled controls. If controls display more restricted immunochemical staining distributions, check additional controls in which one primary antibody is omitted. If these controls display widespread signal overlap, there may be secondary antibody or label cross-reactivity (see *Fig. 2*).

If, having explored all of the above points, difficulties are still experienced in multiple immunochemical staining, it may be advantageous to seek assistance from an on-line immunohistochemistry forum. Several such forums exist, providing advice and guidance on a range of immunohistochemical issues. Three currently popular forums for discussing immunochemical staining methods are:

- Histosearch (http://www.histosearch.com)
- Immunohistochemistry – *In Situ* Hybridization Forum (http://www.immunoportal.com)
- Immunohistochemistry World (http://www.ihcworld.com).

4. REFERENCES

★★ 1. **Krenacs T, Krenacs L & Raffeld M** (1999) *Methods Mol. Bio.* **115**, 223–233. – *A useful review of multiple immunolabeling procedures.*

★★ 2. **van Noorden S** (2002) *Folia Histochem. Cytobiol.* **40**, 121–124. – *A useful review of recent advances in multiple immunolabeling procedures.*

★ 3. **Wessel GM & McClay DR** (1986) J. Histochem. Cytochem. **34**, 703–706. – *Original paper describing the use of unconjugated FAb fragments for blocking and staining (section 1.4.3).*

★ 4. **Franzusoff A, Redding K, Crosby J, Fuller RS & Schekman R** (1991) *J. Cell Biol.* **112**, 27–37. – *Original paper describing the use of unconjugated FAb fragments to alter the effective species of a primary antibody (section 1.4.4).*

★ 5. **Lewis Carl SA, Gillete-Ferguson I & Ferguson DG** (1993) *J. Histochem. Cytochem.* **41**, 1273–1278. – *Original paper describing the use of unconjugated FAb fragments for blocking after the first primary antibody incubation (section 1.4.5).*

6. **Negoescu A, Labat-Moleur F, Lorimier P, *et al.*** (1994) *J. Histochem. Cytochem.* **42**, 433–437.
7. **Vandesande F** (1988) *Acta Histochem. Suppl.* **35**, 107–115.
8. **Jones IW, Barik J, O'Neill MJ & Wonnacott S** (2004) *J. Neurosci. Methods,* **134**, 65–74.
9. **Jones IW & Wonnacott S** (2004) *J. Neurosci.* **24**, 11244–11252.
10. **Akintunde A & Buxton DF** (1992) *J. Neurosci. Methods.* **45**, 15–22.
11. **Soltys BJ, Gupta RS** (1992) *Biochem. Cell Biol.* **70**, 1174–1186.
12. **D'Andrea MR, Rogahn CJ, Damiano BP & Andrade-Gordon P** (1999) *Biotech. Histochem.* **74**, 172–180.
13. **Yi H, Leunissen J, Shi G, Gutekunst C & Hersch S** (2001) *J. Histochem. Cytochem.* **49**, 279–284.
14. **Jones IW, Bolam JP & Wonnacott S** (2001) *J. Comp. Neurol.* **439**, 235–247.
15. **MacIntyre N** (2001) *Br. J. Biomed. Sci.* **58**, 190–196.
16. **Reynolds ES** (1963) *J. Cell Biol.* **17**, 208–212.
17. **Bohn W** (1978) *J. Histochem. Cytochem.* **26**, 293–297.

CHAPTER 6

Confocal microscopy and immunochemistry

Matthew Cuttle

1. INTRODUCTION

Confocal microscopy is a necessary tool for all aspects of the biosciences, as it allows us to image subcellular processes in living cells in real time. By following the methodology laid out in this chapter, it should be possible to produce high-quality images demonstrating genuine co-localization that are free from artifacts and without background fluorescence. Whether confocal microscopy is required and how a confocal microscope works are discussed in this section, followed by details of how to set up a confocal microscope for maximum resolution in section 2. Later sections describe the advantages of multiphoton confocal microscopy and discuss advanced techniques for imaging with multiple fluorochromes and ways of enhancing the final image. These are all components that are important in the study of antigen trafficking, as confocal microscopes are used to produce data that will have a significant impact on human longevity and quality of life.

The principles of scanning confocal microscopy were first described over 50 years ago by Marvin Minsky (1). Bioscience has had to wait for advances in laser and computer technology to make confocal microscopy commercially available and it is now arguably the most rapidly developing area in the optics field. This development is driven in part by the requirements of biological sciences to see molecules in action – to view interactions at the subcellular and even molecular level in both fixed and living cells and tissue.

A confocal microscope is a microscope design where two lenses (one channeling light to the object and one collecting the returning light) are focused on the same point (the object under investigation). Confocal microscopes use epi-illumination, so the excitation light travels down through the objective lens and back up through the same lens to the

Immunohistochemistry: *Methods Express* (S. Renshaw, ed.)

detector. This light path design negates the need for two lenses and produces a system that is inherently confocal (the lens always having the same focus point as itself). By excluding any out-of-focus light, a significant increase in resolving power is obtained – making the technique necessary for research at a subcellular level in biological samples (2). There are several advantages of confocal microscopy over traditional wide-field microscopy techniques. For immunochemical staining, improved resolution and co-localization are two of the most critical.

Confocal microscopy can be put to excellent use in all research fields where traditional immunochemical methods have been used. The increase in resolving power and the ability to demonstrate co-localization of two or more molecules in living cells means that subcellular mechanisms of the inflammatory response, receptor internalization, antigen trafficking, and surface protein expression can be addressed. In fact, imaging the location of a molecule (often relative to other identified molecules) contributes to understanding the pathways surrounding the molecule, leading to discovery of the mechanism of action. The list of potential uses for this technique is therefore quite literally endless.

An example where using confocal microscopy has had a direct impact on our society is in the study of molecule trafficking. Confocal microscopy has been used to gain an understanding of the intracellular sites of antigen processing (3), as fluorescently labeled antigens combined with confocal microscopy allow the visualization of intracellular processing of antigens. Using this methodology, one can identify the mechanisms of antigen trafficking from subcellular compartment to surface presentation and internalization (4). Trafficking of molecules within living cells has been used to demonstrate that virus assembly can be disrupted (5), that leukocyte movement to areas of infection can be improved to increase the efficiency of the human immune response (6), that mechanisms leading to the breakdown of the blood–retina barrier (which can cause blindness) can be elucidated (7, 8), that the molecular signals leading to apoptosis (cell death) in cancerous tissue can be controlled (9), and to determine mechanisms of action for drug treatment of ovarian cancer (10). In a direct practical application, it has also been demonstrated that confocal microscopy linked with the fluorescently labeled antibody KM871 can be used to identify tumors in patients with melanoma (11).

1.1 Do you need to use confocal microscopy?

The confocal microscope is only required if your research question requires subcellular resolution or co-localization of multiple molecules. The theoretical limit of light microscopy resolution can be reached with a correctly set up confocal microscope – a 1.5-fold increase in the resolution of a regular fluorescence microscope that apparently defies the physical laws of optics. This significant improvement in resolution then allows light microscopy to address research questions at the subcellular level that previously required an electron microscope. The advantage that confocal microscopy provides over electron microscopy is that it is a nondestructive technique that can be used on living cells and tissue. Electron microscopy provides ultrahigh resolution, but only on fixed (i.e. nonliving) samples (see Chapter 7, section 1.1). The confocal microscope allows the experimenter to visualize subcellular processes in living cells in real time and has therefore become one of the fundamental tools of cellular bioscience.

A second advantage of confocal microscopy is that it is possible to determine the location of multiple fluorescently tagged molecules in the same preparation without fear of the signal from one molecule contributing to the others. It is now possible to determine interactions at the molecular level of four or more molecules simultaneously in the same preparation.

Contrary to how one might feel when watching an expert user operating a confocal microscope, however, there is no magic involved in producing an image on the computer screen. Spells and incantations can be bypassed by learning simple rules for adjusting the various knobs and sliders one is faced with when setting up for the first time.

1.2 How does the confocal microscope work?

There are a number of different methodologies for achieving a confocal image, but in principle the different manufacturers use a similar optical light path (see *Fig. 1a* and *b*, also available in the color section) as originally patented by Minsky over 50 years ago. A light source is focused on to the sample through the objective lens, and the fluorescence generated is collected back through the lens, appropriate filters, a pinhole, and on to a detector. The key to achieving confocality is in the pinhole. The light source and the detector are both focused on the same point, and any fluorescence generated from out-of-focus material (material above and below the plane of focus) does not make it through the pinhole and on to the detector (see *Fig. 1a* and *b*).

The light path described above generates a single confocal pixel. To generate the kind of images we recognize as confocal, the excitation light is scanned across the sample in *x* and *y* directions to produce a two-dimensional array of confocal picture elements. Three-dimensional image stacks can then be created by simply moving up and down the focal plane, taking an image at each *z* position (see *Fig. 1c*, also available in the color section).

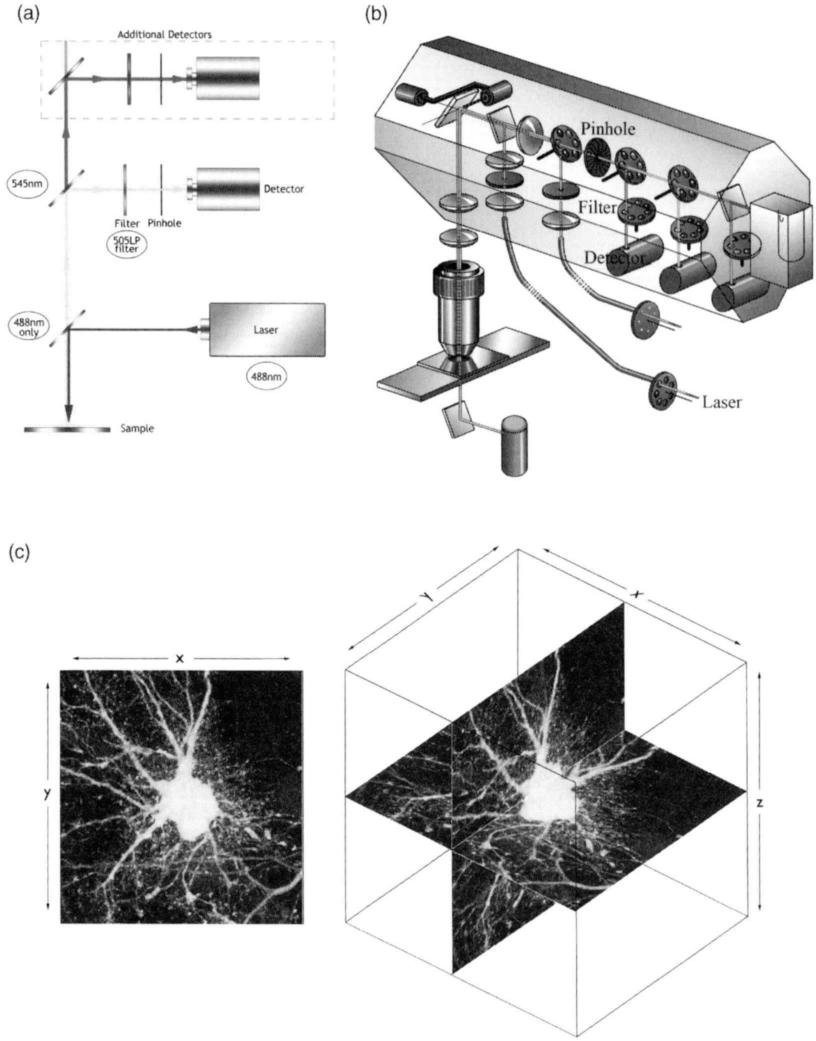

Figure 1. The inner workings of a confocal microscope needed to produce a three-dimensional reconstruction of the object (see page xvi for color version).

(*a*) The basic light path of a confocal microscope, with example filters for GFP (green fluorescent protein) imaging. The path starts with the laser light source (488 nm blue laser) and is reflected down to the sample with a dichroic mirror (488 nm only). Green fluorescence generated in the sample goes up and through the first dichroic mirror, is reflected off a second dichroic mirror (545 nm), through a filter (505 nm long pass (LP)), a pinhole and on to the photomultiplier tube detector. Additional detector(s) for longer-wavelength light can be added into the system beyond the second dichroic mirror, as shown. (*b*) A typical confocal light path inside a traditional, filter-based Leica microscope. The laser line, dichroic mirrors, filter wheels, and detectors can be seen, showing a light path similar to (*a*). In this example the pinhole is not directly in front of the detectors, but in front of the filter wheels. This configuration makes little practical difference to the system. (*c*) *x*, *y*, and *z* planes provide coordinates in a three-dimensional system, whereas two-dimensional systems (i.e. regular light microscopes) only demonstrate *x* and *y* planes.

2. METHODS AND APPROACHES

The following sections describe how to set up a confocal microscope for maximum resolution.

2.1 Selecting fluorescent dyes

Before attempting any fluorescence-based immunochemical staining, you should be aware of the excitation and emission properties of the fluorochromes you intend to use (see Chapter 3, *Table 2*) and then check that your microscope has the appropriate laser lines and filter blocks for your requirements. There are a number of standard laser lines that are available on most systems (generated by argon, krypton, helium–neon, and more recently diode lasers). You may also have available an ultraviolet (UV) laser, but these are less common in the biosciences, as UV light can cause significant photodamage to living cells and tissue. They are, however, very useful for imaging the nuclear stain DAPI, which, contrary to popular belief, works well on many live-cell preparations. Another option is a 'far-red' titanium–sapphire laser for multiphoton confocal microscopy. This latter system has a number of advantages over normal confocal microscopy and is discussed below (see section 2.8).

Classically, the ideal scenario for multiple staining in one sample is to have the two or more fluorochromes excited by different laser lines. A typical quadruple labeling experiment may contain the blue dye DAPI, excited by the UV laser 364 nm line, a green dye such as fluorescein isothiocyanate (FITC) or green fluorescent protein, excited by the argon 488 nm line, a red dye such as rhodamine, excited by the helium–neon 543 nm line, and a far-red dye such as Cy5, excited by the helium–neon 633 nm line. In this example, it is possible to scan the sample sequentially with each laser so that at any one time only one of the fluorochromes is excited. This guarantees that for each of the four images collected there can be no cross-talk or bleed-through (where one dye signal appears on another light path), as at any one time only one fluorescent molecule is excited. This ideal is not always possible and you may be forced to choose fluorochromes with similar excitation properties. Under these conditions, the larger the difference between the emission spectra of the dyes, the better. With a large difference in emission spectra, it will be possible to separate two signals with appropriate filters and dichroic mirrors. One must then run parallel samples, however, stained with only one of the two fluorochromes. These must then be imaged under identical conditions to your original experiment to demonstrate that there is no bleed-through, proving that any co-localized signal is genuine.

Co-localization studies of multiple dyes with similar, or even identical, excitation and emission spectra are possible and actually have a number of advantages over the classic sequential excitation method. This is discussed later in this chapter in more detail.

2.2 Setting up the light path on the microscope

Setting up the light path is easy if you follow the path of the light from its source to the detector. If, for example, you need to set up an FITC light path (excitation 488 nm, emission 500–530 nm), start at the laser (see *Fig. 2*, available in the color section, page xvii). Select the 488 nm line of the argon laser and follow it to the first dichroic mirror (an optical component that reflects light with short wavelengths, but allows through light with longer wavelengths). If FITC is the only fluorochrome in the sample, choose the 488 only dichroic, as this will allow all wavelengths of light longer than 488 nm back to the detector. It is important to note that selecting a different dichroic mirror at this point in the light path will change the spectrum of light that can reach the detectors. If you choose a 488/514 dichroic, this means that wavelengths of light around 514 nm cannot reach the detectors and your FITC sample will appear weaker. In a study where you want to compare relative intensities, you must maintain identical light path settings and this includes the dichroic mirrors.

Once you have chosen the correct dichroic mirror (one that allows all laser lines through and the maximum fluorescent signal back to the detectors), follow the light path down to the sample. Fluorescent light from the sample then returns back through the objective lens and reaches the dichroic mirror you have already selected. Emitted light must be of longer wavelength than the excitation light in accordance with Stokes' law, so light longer than 488 nm wavelength will pass through the dichroic and on to the detector.

In systems with two or more detectors, additional dichroic mirrors will be in the light path. The majority of systems will have three or sometimes four detectors. This allows simultaneous acquisition of multiple fluorochrome signals and speeds up data acquisition. Select the dichroic mirror that sends the maximum amount of fluorescent light to a single detector. In the FITC example, a 545 dichroic will reflect all wavelengths of light lower than 545 nm to one detector. Under the current settings, only light between 488 and 545 nm will reach the detector.

To reduce further any possibility of nonspecific fluorescence contributing to the signal, a filter is placed in the light path in front of the detector. On a confocal microscope, there are two types of commonly used filters: long pass and band pass (short-pass filters are also used in multiphoton confocal microscopy). Long-pass filters allow all light longer than the specified wavelength through, whereas band-pass filters allow through wavelengths of light between two specified points. In the example, if we choose a 505–530 nm band-pass filter, we reduce the likelihood of nonspecific fluorescence contributing to the image, but in a sample with a good FITC signal, a 505 nm long-pass filter would be equally good.

An alternative solution to mirrors and filters exists in confocal design. There are a number of optical problems associated with passing light through glass. For example, although a 488 dichroic allows through all wavelengths of light longer than 488 nm, these wavelengths do not pass through with equal efficiency. As a general rule, approximately 4% of your fluorescence signal is lost each time the light passes through a filter or is reflected off a mirror. Electronic tunable filters can be used to solve these problems and can result in fluorescent signals up to 40% stronger at the detector. This may be an important issue when your samples have weak fluorescence. The general principles of setting up the light path for each fluorochrome still apply, even in a system with electronically adjustable filters.

In summary, follow the light path from laser to detector for each of the fluorochromes in your sample. Set up the shorter-wavelength light paths first, reflecting the emitted light to the first detector. This methodology will avoid any conflicts in dichroic mirror settings (if your system has them) as you introduce the longer-wavelength light paths. If you have more fluorochromes than detectors, it is possible to reset the light path so that the same detector can be used for multiple fluorochromes, but this can be a slow process (moving filters and mirrors takes time) and speed can be important in live-cell imaging or when your fluorochrome is one that photobleaches (fading of the sample due to the destruction of the fluorescent molecule from the light source).

2.3 Choosing the right lens

An important factor in lens choice in confocal microscopy is getting the *z* resolution that you require. Lower-power lenses will give you a wide field of view, but the loss of resolution in *z* is significant. A 10× lens with a numerical aperture (NA, the light gathering capacity of the lens) of 0.3 has a *z* resolution of only 45 μm, comparable to a regular light microscope image. Using these settings, it is unlikely that you need a confocal microscope, as a basic fluorescence system will generate a very similar image. Low-power confocal images are usually generated by creating a montage of higher-resolution pictures. This is a feature normally integral to the software of a confocal equipped with a motorized *xy* stage.

The *z* resolution of a confocal image is determined by the NA of the lens you use (see *Table 1*). For maximum resolution, choose a 63× oil-

Table 1. Maximum optical slice depth available for different objective lenses
A lens should not only be selected on magnification and light-gathering capacity (numerical aperture (NA)), but on the *z* resolution required for the experiment. Maximum optical slices are given for UV, blue, and green excitation wavelengths.

			Maximum optical slice		
Medium	**Magnification**	**Typical NA**	**UV**	**Blue**	**Green**
Air	5×	0.15	38.5	46.4	56.5
Air	10×	0.3	9.6	11.5	14.0
Air	20×	0.5	3.4	4.1	5.0
Air	40×	0.7	1.5	1.8	2.1
Water	20×	0.5	4.6	5.6	6.7
Water	40×	0.8	1.7	2.1	2.6
Water	63×	0.95	1.2	1.5	1.8
Oil	40×	1.3	0.7	0.9	1.1
Oil	63×	1.4	0.6	0.7	0.9
Oil	100×	1.3	0.7	0.9	1.1

immersion objective (NA of ~1.4) where possible, which will result in an optical section close to 0.5 μm thick. It may appear that the choice is simple, to always use the 63× oil-immersion lens. However, imagine if your sample is 100 μm thick. You would need to take at least 200 images (in practice 400 images; see section 2.7) to collect the total fluorescence signal and often this is not practical.

Typically, lenses with a higher magnification than 63× have a lower NA and have a lower *z* resolution (although recent developments have produced very high-resolution lenses with equally high NAs, for use with total internal reflection fluorescence microscopy). Given that the confocal microscope will invariably have a digital zoom function that retains the number of pixels in the image (and even an optical zoom on some systems), it is not necessary to collect images with a lens magnification higher then 63×.

There are additional ways to improve the *z* resolution of your image by collecting a *z* stack. This is covered in detail in section 2.7 later in this chapter.

2.4 How to set the digital image settings

Another determinant of the final image quality is the digital image settings you choose. There are two components, the number of pixels in the image and the bit depth of each pixel.

Standard confocal images contain 262 144 pixels (512 × 512). At this resolution the image will begin to appear pixilated if printed out larger than US letter-sized paper (A4 paper for UK readers). For most research questions this is more than adequate, but higher resolution may be necessary for large high-quality printouts. In theory, collecting an image at 1024 × 1024 pixels should appear no different on the computer screen in front of you, as both images contain more pixels than can be discriminated by the human eye. In practice, the 1024 × 1024 image will almost always appear better (see *Fig. 3*, available in the color section, page xviii). This apparent contradiction between theory and practice is the result of the fact that the 1024 × 1024 image has four pixels for every single pixel in the 512 × 512 image. The increase in image quality of the higher-resolution picture is simply the result of averaging of the four pixels into one screen pixel. If the lower-resolution image is taken three more times and an average generated, the image will be identical to the larger resolution image – but the image file size will be four times smaller. File size is not an issue for single images, but a *z* stack of 50 or more images becomes a significantly large file so that even simple image analysis becomes a time-consuming, labor-intensive task.

The bit depth of an image is the number of grayscale levels that can be represented at each pixel, i.e. the number of different grays in the image, the darkest being black and the brightest white. The usual choices are 8-bit or 12 bit, although 16-bit image depth may also be available. In an 8-bit image, there are 256 possible gray levels (i.e. from 2^0 to 2^8 (or 00000000 up to 11111111 in binary) is 0 to 255 in decimal). In a 12-bit (2^{12}) image, 4096 gray levels can be encoded. To put this in perspective, the human eye can distinguish about 16–18 gray levels. This being the case, there is absolutely no difference in appearance between 8-bit and 12-bit images. Twelve-bit images generate file sizes twice as large as 8-bit images, so should be used only when necessary. Twelve-bit image depth is necessary when looking for small changes in fluorescence (significantly less than 10%), as small signal changes in 8-bit-encoded images can become lost in the background noise. A 1% change in fluorescence is a change of two or three grayscale levels in 8-bit images, but a change of more than 40 grayscale levels in 12-bit images.

In summary, 512 × 512 pixels, 8-bit images are ideal for most experiments and generate small file sizes that save time during analysis. It is important, however, to collect larger-resolution images for large printouts and 12-bit images if you are looking for small changes in fluorescence in your sample.

2.5 Getting ready to scan

Finally, after selecting fluorochromes, lens, and digital image settings, you can press the scan button on the confocal. Well, almost. Set up the basic microscope settings first; find and focus your object of interest, focus and centre the condenser, and set the field iris correctly. These are standard settings on a light microscope and are not specific for confocal microscopy. Although the latter two settings make no difference to a fluorescent image (they are not in the light path), they are vital to obtaining a good bright-field/transmission image (an image that appears as a regular light microscope image that is generated by collecting the laser light that travels through the sample) and will come into effect if the confocal microscope has a transmitted light detector (a detector that produces an image using the laser light that travels through the sample).

When starting with a new sample on the confocal, turn all of the detector gains and offsets to zero. Constant scanning with a signal bright enough to saturate the detector will eventually burn out the detector (literally!). This can be avoided by turning the detector gains down to zero when first scanning your object.

2.6 Optimizing the image collection settings

Set the pinhole(s) in front of the detectors to 1 Airy disc unit. The pinhole diameter determines the *z* resolution of your image (the best *z* resolution is determined by your choice of lens, but you can choose to set to a lower resolution). When light passes through the optics of the system, light from a single point source is blurred slightly, producing an image of a disc rather than a point. This is the Airy disc. By setting the pinhole to 1 Airy disc unit, you are collecting all of the available light from each point of the image and no more. Although it is possible to generate a *z* resolution slightly higher than that defined by 1 Airy disc (by closing the pinhole slightly), in most cases this should be avoided, as you are throwing away fluorescent signal strength with very little gain in resolution. If the pinhole is opened, on the other hand, this significantly increases the brightness of the signal, as more fluorescence is detected. At the same time, resolution is lost, as the system is now detecting out-of-focus light. Opening the pinhole is a significant temptation when setting up a system, as this is the easiest way to generate an image, but by doing so you are throwing away the very resolution that you paid for by purchasing a confocal microscope.

The next step in obtaining a high-quality image may appear to have a certain amount of 'juggling' involved. Laser power, scan speed, detector gain, detector offset, and pinhole diameter all have consequences on the final image quality and adjusting any one of the settings has an impact on all of the others. For example, increasing the laser power would necessitate lowering detector gain, increasing detector offset, increasing scan speed, or reducing pinhole diameter. There are, however, good starting points for each of the settings and a reasonable order to adjust each setting.

After setting the pinhole, select an appropriate laser power. The choice of laser power is always a compromise: the higher the laser power, the greater the fluorescence (more light in means more light out), but higher laser powers can induce photobleaching (more light in means more photodamage). Given the trade-off between fluorescence signal strength and photobleaching, *Table 2* provides a list of common lasers available, their power, and a good starting point for the percentage maximum power of the laser you should use. These are a guideline only and will have to be adjusted depending on the properties of your sample.

With the lasers on, and detector gains and offsets set to zero, you can at last press the scan button and what you should get is a black image. A lot of effort for a black screen, but an experienced user never panics when faced with blackness. Gradually turn up the detector gain until an image is obtained. For a sample with a good fluorescence signal, there should be no problem at this point. If no signal appears, however, increase either the laser power or slow down the scan speed. The scan speed determines the amount of time the laser spends generating each pixel in the image. Slower scan speeds increase laser exposure time for each pixel, so this is in effect another way of increasing laser power. As a last resort, you can open the detector pinhole. This will significantly increase the amount of fluorescence returning to the detector, but remember that you are trading off resolution to get this increase in fluorescence.

The detector offset is a way of setting the zero point (black level) in your image and can be used to remove background fluorescence. There are two approaches to setting the gain and offset of a detector. If you require an image that is simply a representative picture and you do not intend to

Table 2. Typical laser wavelengths and approximate starting values for imaging biological samples

Laser wavelength (nm)	Power (mW)	Optimal starting power (%)
UV 351	80	1–3
UV 364	80	1–3
Blue 458	20	5–10
Blue 477	20	5–10
Blue 488	20	5–10
Cyan 514	20	5–10
Green 543	1.5	50–100
Yellow 561	20	5–10
Red 594	5	15–25
Far red 633	5	15–25

perform any image quantification, then simply adjust the offset and gain on the detector until the image appears as you want it to be in its final form. Although this is not a recommended method of image generation, it is mentioned here, as optimal image settings often result in an image that appears dim to the eye and will need adjustments in brightness and contrast for presentation.

Optimal settings for gain and offset on a detector are obtained using a suitable color look-up table on your system. The color look-up table determines how the signal at each pixel is represented, e.g. a faint signal may be represented as dark blue, a brighter signal as green, and the brightest signals as red (this would be a typical 'rainbow' look-up table). An excellent choice of color look-up table for determining the correct gain and offset settings is the 'range indicator' (Zeiss) or 'glow over/under' (Leica). All manufacturers provide an equivalent. This look-up table represents zero photons as one bright color (often blue or green), maximum photons (a saturated signal) as a contrasting color (often red) and all other intensity levels in between as more muted colors. The gain on the detector should be increased until saturated pixels appear and then turned back down until they just disappear. The offset in the image should then be adjusted in the same way, turned down until the majority (but not all) of the zero pixel values have gone. At this point, the range of intensities in your sample utilizes the full dynamic range of the detector and maximum information about the image is encoded. Following this method will always allow you to produce the best-quality image, as it collects all of the available information from the sample. This optimal image may appear somewhat fainter than you would choose simply by eye, but enhancements in brightness and contrast can easily be achieved offline to produce the final image you need. If you collect images without setting the gain and offset optimally, you will not collect all of the information available from your sample and will ultimately have relatively poor-quality images.

For applications where fluorescence intensities between samples are to be compared, it is best to leave the detector offset at zero (where zero photons of light is represented as zero in the image), because only in this position does the scale represent an actual measure of the light coming back from the sample. It is also a good idea to set the gain lower than optimal. If future samples have an increase in fluorescence, the gain would have to be reduced. All previous samples would then have to be re-imaged under the new detector settings, as the gain on a photomultiplier tube is never exactly linear and so changes in gain cannot be realistically corrected for post-imaging.

By setting pinhole diameter, laser power, gain, and offset in that order, there should be little need for 'juggling' of the controls, although some final adjustments may be required, if only to satisfy the experimenter that the best image has been obtained.

Generally, the more time you spend generating an image, the better the quality of the final image. This rule breaks down if the sample is prone to photobleaching. For samples relatively immune to photobleaching, slow scan speeds and image averaging are recommended. Even averaging of two images significantly reduces background 'speckle'. If time permits, averaging of four images produces an excellent quality image.

2.7 Optimizing *z* stack image collection

Confocal microscopy is often used as a method of producing high-quality three-dimensional image stacks, allowing localization of fluorescence within a sample. Once you have obtained an appropriate single image with optimal settings, it is simply a matter of focusing through the tissue and collecting a series of images. There are a number of things to watch out for at this point. First, the detector gain and offset have been set for a single section and it may not be the section with the strongest fluorescent signal. As you focus up and down the sample, you may need to readjust the detector settings.

Once you have established the *z* range of your sample (i.e. where the first and last image are to be recorded), you must then choose how many images to take to make up the *z* stack. The absolute minimum difference between each optical slice is defined by slice thickness. If you are collecting optical slices of 2 μm in depth, then you could sample through the *z* stack every 2 μm and collect all of the available fluorescence. If you collect at a larger *z* interval, then your image stack will be streaked with black planes where you simply did not sample the object.

Minimum sampling without any overlap between optical slices is significantly prone to artifacts, however. If you require a *z* resolution of 2 μm, you must sample through the *z* stack at 1 μm intervals (i.e. half the required resolution). This is known as the Nyquist sampling limit. Sampling at less than the Nyquist limit can lead to aliasing in the image. In an imaging sense, aliasing translates as things appearing closer together or further apart than they really are. Imagine in this example with 2 μm sections that you are picturing two point sources 2 μm apart: it is just possible that they will appear in the same section and appear co-localized, or in consecutive sections, or in two sections with an empty section between them, i.e. with 2 μm slices it is only possible to locate an object in *z* to the nearest 4 μm. By introducing an overlap between sections, this problem can be avoided. Typically, a 50% overlap is recommended, as this maintains the *z* resolution obtained in each section (i.e. 2 μm sections should be taken every 1 μm, giving a 2 μm *z* resolution).

If sampling a *z* stack with a 50% overlap maintains the *z* resolution, what if a larger overlap is introduced? As you may imagine, this improves the *z* resolution further, but only up to a point. The quality of the sample, the NA of the lens, and the wavelength of light all have an upper limit. It is difficult to get beyond 500 nm *z* resolution with a confocal microscope. In practice, your *z* stack resolution is more often limited by time and patience, as a *z* stack that takes 5 min to collect with 4 μm resolution may take well over 2.5 h to collect with *z* resolution approaching the theoretical maximum and your sample may photobleach before the stack is complete!

There is an additional adjustment that can be utilized on large *z* stacks. If you are working with relatively thick tissue slices (greater than 50 μm), as you travel deeper into the slice the fluorescence signal appears to fade. This loss of signal results from the fact that it is impossible to get light to travel through relatively solid objects without being reflected or refracted in some way. The laser light traveling into the sample, and to a larger extent the fluorescence traveling back from the sample, is more likely to be deflected if it has more tissue to travel through. As the confocal microscope is designed to collect only in-focus light, any deflected light is lost to the final image. (This also explains why some samples that appear bright under wide-field fluorescence microscopy can appear only weakly fluorescent under a confocal microscope.) In order to compensate for fluorescent signal loss in thick samples, it is possible to introduce a linear increase in either the detector gain or the laser power (or both) as you sample deeper and deeper into the tissue. This can be a useful technique to produce a series of images that penetrates the tissue deeper than previously thought possible, but this linear 'correction factor' cannot be used when image quantification is required. The loss of fluorescence from deep tissue imaging is not a linear function of depth traveled into the tissue; it is very definitely unpredictably nonlinear. Comparisons of fluorescence signals within or between samples should be avoided if this technique is used.

2.8 The advantages of multiphoton confocal microscopy

For true confocal imaging, and collection of *z* stacks that penetrate deep into tissue, a multiphoton confocal microscope is a distinct advantage over traditional confocal microscopy. The multiphoton laser is a far-red laser (usually tunable from around 700 to 1000 nm). Longer-wavelength light penetrates deeper into objects than shorter-wavelength light. How then, can a far-red laser at wavelengths of more than 700 nm be used to excite fluorochromes with UV and visible light excitation spectra, apparently defying Stokes' law? In simple terms, if two photons (that is, multiple photons) from light at 700 nm arrive at a UV-excitable fluorochrome simultaneously, the photons interfere with one another to produce an excitation at approximately half of the original wavelength, i.e. 350 nm. In order to achieve this multiphoton effect, powerful titanium–sapphire lasers are pulsed in a way that focuses large numbers of photons into a small period of time (100 fs to 1 ps). This gives the confocal system the depth penetration of a far-red light source, but the ability to excite UV and visible fluorophores.

A second advantage of the multiphoton lasers is that it is inherently confocal and requires no pinhole. Only the in-focus laser light contains enough power to excite your fluorochrome, and laser light above and below the plane of focus is not able to generate fluorescence. This means that all fluorescence generated can be collected without passing through any mirrors, filters, or a pinhole, significantly increasing the signal strength in the image (remember that approximately 4% of the signal is lost for every optical component in the light path).

The third advantage of multiphoton confocal microscopy is that far-red light is less damaging to biological tissue samples than visible and UV light. The multiphoton laser therefore generates less photobleaching and less photodamage (unless photobleaching is required, as in techniques such as fluorescence recovery after photobleaching (FRAP), where the fluorochrome is deliberately bleached away, so that the movement of new fluorescent molecules into the area can be measured, in which case the power of the multiphoton laser proves useful). In practice, this means that even stronger laser light can be used to penetrate deeper into the tissue.

Finally, most biological samples will autofluoresce under UV, blue, or green excitation wavelengths (especially plant materials), and using the far-red laser can significantly reduce the autofluorescence signal in your samples. Autofluorescence is the inherent ability of tissues to fluoresce, a problem familiar to anyone who works with biological samples.

The combination of more-penetrating wavelengths, inherent confocal imaging, and less damage to biological samples makes multiphoton imaging an excellent technique for visualizing subcellular processes in thick tissue samples.

2.9 Advanced techniques for imaging with multiple fluorochromes

There are two further approaches to imaging with multiple fluorochromes that are pertinent to co-localization in immunochemical staining. Often it can be difficult to obtain two fluorochromes with distinctly separate excitation and emission spectra (see Chapter 3, *Table 2*). Certain dyes have broad emission spectra (propidium iodide, for example) and will contribute to images generated through blue, green, and red filter sets. However, all is not lost at this point, as there are a number of ways to separate two dyes with overlapping emission spectra.

2.9.1 Separating fluorochromes by their relative intensities in two channels

Perhaps the simplest methodology for removal of contaminating signals in co-localization studies is to compare relative intensities in two channels. This methodology works when only two fluorochromes are involved. For example, if 10% of the signal from the green fluorochrome is contributing to the red image, then this constant can be removed from the red channel by simple image subtraction. (This percentage can be calculated by looking at the red image in a position that is known from the morphology of your sample to have green signal present but no red signal present, i.e. if it is a red nuclear dye, look for red signal in cytoplasmic compartments.) Any residual co-localization of the red and green signals after this mathematical process should be genuine co-localization. This process works equally well when the fluorochromes involved contribute to both images generated, as long as neither signal is weak. The difficulty in this method is that one needs to identify in the image an area that is predominantly or preferably purely a green signal and predominantly a red signal – an easy task if the fluorochromes involved have specific morphological targets, but not so easy if your expected result is significant co-localization.

2.9.2 Separating fluorochromes by their emission spectra (and removing autofluorescence)

When signals are separated by their relative intensities, each pixel in the image provides two measurements (intensity in red and intensity in green, for example). However, more information about the fluorochromes is available. There are confocal systems that collect the spectral emission profile of each pixel in the image. Some designs, like the Leica SP2 series, inherently allow collection of emission profile data. Other systems may require this to be a specified option.

In order to obtain spectral profile information, the sample is imaged a number of times and the emitted light returning to the detector is filtered by a narrow-band-pass filter (around 10 nm works well; on a Zeiss Meta this window is fixed at 11 nm, but on the Leica system this can be adjusted to suit the sample). This filter is moved gradually through the wavelengths of the emission spectra of the sample. Consecutive images in the generated stack represent different emission wavelengths of light. A lambda stack (λ being the wavelength of light) of images means that each pixel in the final image contains spectral emission information. In order to use this information, the ideal experimental design is to make up multiple samples, each of which contains only one fluorochrome. These samples are then imaged to produce a spectral emission profile for each fluorochrome. In the final experiment, a sample containing all of the fluorochromes is imaged and the spectral profiles of the individual fluorochromes are used to 'unmix' the lambda stack into its multiple fluorochrome components. This method of separating fluorescent signals is also an excellent method for removing autofluorescence from a sample. Autofluorescence has a spectral profile and can be separated from genuine signals by spectral unmixing, similar to intentionally added fluorescent signals.

Although it is advisable to build up an emission spectra database for your confocal microscope, it should be noted that the experimental process that individual samples go through will alter the samples' spectral profiles – enough to invalidate this method. Tissue fixation can significantly change the autofluorescent properties of your sample, for example (see Chapter 9, section 2.8.7). It is advisable to generate new spectral profiles for different tissues, or for tissues that have been processed differently.

In reading this section, you should get an appreciation of the importance of control experiments (see Chapter 9). You should be aware that with enough laser power and a sensitive photomultiplier tube detector system, it is possible to get just about anything to fluoresce. Cells will happily fluoresce at FITC emission wavelengths, for example. Control experiments are critical, perhaps even more so when co-localization is generated through mathematical manipulation of the image. You can ensure that your signal is genuine by testing untreated samples under identical experimental conditions. It is all too easy to generate autofluorescence signals that one accepts as genuine simply because the signal appears in the correct place in the sample. Don't be fooled that just because your image looks right, it therefore must be genuine.

2.9.3 Co-localization studies with fluorochromes with identical excitation and emission spectra

Fluorochrome signals can be separated by their excitation wavelengths, filtering, relative intensities, and spectral profile unmixing. But what if the excitation and emission spectra of your fluorochromes are identical? Perhaps before this question is addressed, it may be appropriate to ask the question, why would you use two identical fluorochromes in your sample, especially in co-localization studies?

In fact, there are a number of reasons why it is advantageous to use identical fluorochromes. Anyone familiar with fluorescent staining will be all too aware that, regardless of the experimental set-up, the green fluorescent signal is generally better than the red. This phenomenon is caused by the chemical nature of the fluorochromes. It is possible to fit more functional green fluorescent subunits on to a molecule than red ones. As more fluorescent subunits are added to each molecule, they tend to interfere with one another such that the fluorescence generated is quenched. With a green fluorescent subunit, up to eight fluorescent subunits can be added to a single fluorochrome before the molecule 'autoquenches', but with a red fluorescent protein, only three or four subunits can be used. This makes green fluorochromes at least twice as bright as the equivalent red fluorochrome. As well as green fluorochromes having more fluorescence, the blue lasers that excite green fluorochromes (argon or krypton–argon) have significantly more power (around 20 mW) than the equivalent helium–neon lasers used to excite red fluorochromes (1–1.5 mW).

Both fluorescence intensity and laser power can, within reason, be adjusted so that green and red fluorochromes appear equally bright on a confocal system. However, the limit to resolution of a confocal microscope is limited by the NA of a lens (its light-gathering capacity), which in turn is defined by the wavelength of light passing through the system. Shorter-wavelength light has a smaller Airy disc and therefore better resolution. In an optimally adjusted confocal system, a pure point source of green fluorochrome has an Airy disc close to 0.7 μm and produces a three-dimensional volume in a confocal system of around 0.07 μm^3, whereas longer-wavelength red light with an Airy disc of just over 1 μm would result in a volume of 0.14 μm^3. Although the dimensions involved are small, if it is critical for your experiment to show true co-localization, this is only possible using fluorochromes with identical excitation wavelengths. It makes sense then, if possible, to use green fluorochromes with identical excitation and emission properties such that they follow the same light path through the optics of a system. Only then could your two signals be considered co-localized at close to the molecular level.

Having solved the problem of how to set up a confocal microscope to show error-free co-localization, we now have to solve the less-than-trivial problem of separating the two identical signals in a meaningful manner. Fortunately, this has been done for us. There are now three green dyes available for green/green (or even green/green/green) co-localization studies. All dyes have identical excitation (488 nm) and emission (505–530 nm) wavelengths, but differ in their fading properties. Using these dyes for high-resolution co-localization studies takes advantage of the fact that the more recently developed dyes (Alexa Fluor 488, for example) are less susceptible to photobleaching than the more traditional dyes (FITC). As stated earlier in the text, there are up to eight possible fluorescent subunits on a green fluorescent molecule. The three new dyes have either eight FITC subunits (which fade rapidly), eight Alexa Fluor subunits (which are fade resistant), or a 50:50 mixture of the two units (which fades partially). The dyes are separated by their fading properties. In practice, this requires repeated imaging of the sample over time. The initial image will contain all three green dyes, but as the sample is repeatedly exposed to the blue laser, the fading properties of each pixel in the imaged are determined by the relative amounts of each dye at that point. Once the time series is generated, it becomes a relatively simple mathematical operation to extract the three different dye signals. Put simply, the first image contains all three dyes, later images contain only two dyes (the FITC-based dye has faded) and the final image contains only one dye (the pure Alexa Fluor dye). Using image subtraction, all three dye signals can be separated.

One can foresee a few limitations of this technique; perhaps the most likely is that taking *z* sections of your samples may be difficult. With each slice, the rapidly fading FITC-based dye has the potential to fade as the material is repeatedly exposed to the blue laser with each image slice. This is another area where multiphoton fluorescence microscopy would be advantageous, however, as in this case the out-of-focus material in the sample is exposed to nondamaging (and more importantly nonphotobleaching) far-red light.

With recent developments in confocal microscopy, one can now separate multiple fluorochromes using their excitation wavelengths, appropriate dichroic mirrors and emission filters (either fixed or tunable), and emission spectra (even when overlapping), and now even 'identical' dyes can be separated by their photobleaching properties. Within each of the sections above there is a small guide to the limitations of each method and it is hoped that you will now be able to make an informed choice as to the best approach for your particular requirements.

2.10 Enhancing the final image

Once you have created your image, there are a number of ways that the information in it can be enhanced. This section only covers a few basic elements of image enhancement that should be used by everyone prior to publication. Some care is required in your choice of manipulation, as there reaches a point where the original image may as well be uploaded to Adobe Photoshop and elements added or removed at will to produce the image you want. Historically, a picture is used instead of a thousand words, but be careful not to commit scientific fraud inadvertently by excessive manipulation of data.

2.10.1 Choosing an appropriate image gamma and color look-up table

In a confocal image, you have a digital representation of an object. In standard imaging techniques, an image is represented by its genuine color and its genuine intensity, so that a green dye at half maximum brightness is represented as a green pixel at intensity 128 (in an 8-bit image with 256 gray levels). There is no requirement to represent the true color or to have a linear representation of intensity, however. If it was necessary to draw attention to the brightest pixels in the image, for example, one could chose to represent all pixels with values of less than 128 by zero intensity (i.e. making all the weakly fluorescent parts of the image black). The relationship between actual pixel intensity and that represented in the image is known as the image gamma. If the image gamma is 1, then the relationship is linear; a value of less than 1 enhances the brighter pixels and a value of greater than 1 brings up the darker pixels. Changing image gamma is a perfectly acceptable practice and is no different from changing the brightness or contrast values of your image. Any manipulation of image properties must be applied equally to both control and experimental images.

As stated above, there is no requirement to represent a green dye with green pixels. Color can be used to enhance the regional differences in the image, with low-intensity pixels represented as blue and high-intensity pixels represented by red or white pixels (often referred to as a 'rainbow' look-up table). It is possible to design your own look-up table that will specifically enhance the features you need to make clear, but the color look-up table must always be presented with the image. It is even possible to represent the brightest pixels as black and the darkest pixels as black, and represent the mid-range pixels as being the brightest – possible, but definitely not recommended.

2.10.2 Image filtering

Images can be cleaned up in all offline analysis packages using image filters. Low-pass filtering will remove any background noise (dots and spots) from an image, but will result in a slightly out-of-focus look to confocal images, as all edges are smoothed. High-pass filtering enhances edges, but enhances the background dots and spots. There is now a vast array of different filters that can be applied during image analysis and the best advice for selecting filters is to try them out and see whether you like the result. One filter is worth mentioning as worthwhile trying in all cases, however. 'Top Hat' filtering removes the speckle from an image without smoothing the edges of larger objects. The name of the filter suggests how it works. The filter is passed over the image and any items that fall into the hat (small dots and spots) are removed from the image. Larger objects that do not fit in the hat are left unaltered.

2.10.3 Image deconvolution

Changing the gamma, look-up table, and filtering can only remove data from an image, with the aim being to leave only the data that you want. Deconvolution can actually put back data that has been lost during image collection. This technique can be used on any image, but especially applies to confocal microscopy, as the data loss in a confocal system is remarkably predictable.

All optical systems lose high-resolution data. A single point source of light, when passed through a lens, comes out as a blurred (Airy) disc, effectively low-pass filtering the image. Neighboring points in an image have overlapping Airy discs, further contributing to image blur. In three dimensions, the volume of blur in a confocal system is hour-glass shaped, with the narrow neck of the glass being in the plane of focus. This blurring of the light is known as the point spread function (PSF) of the system. Each pixel in a confocal image has an identical PSF, and in a *z* stack each slice has a component of the PSFs from the pixels above and below (in a single photon set-up, at least), as well as the pixels within the slice. PSFs are 'almost' independent of position in the image (small variations can be introduced by spherical aberrations for example, but these can be ignored). Under these circumstances, it is possible to remove this predictable blurring of the image and restore the high-frequency information.

Using the confocal microscope to image brightly fluorescent beads that are beyond the resolution of light microscopy, it is possible to measure the PSF on your microscope. Fluorescently labeled 200 nm diameter beads are commercially available and are effectively a point source of light (being beyond the resolution of the system), so the image you obtain from a bead is the Airy disc (blur) on your microscope. There are a number of three-dimensional deconvolution packages available, each using different mathematical algorithms to reconstruct high-resolution data within the image. This is no simple task, as when your original raw data (the fluorescence from your sample) is convolved with the PSF of the system there are a large number of points in the image that are multiplied by zero. (This is because for each pixel in the image the blur from all of the other pixels has to be factored in, and the majority of pixels in an image are too far away to make any contribution, hence they provide a zero in the calculations.) It is not possible, therefore, simply to reverse the operation, as divisions by zero tend to upset computers and mathematicians. Although three-dimensional deconvolution software packages are expensive, it is generally recommended that all confocal images are deconvolved. Deconvolution packages come supplied with standard PSFs, but it is important to be able to use a PSF measured on your microscope to achieve the best results, so make sure that the software you choose has this option.

The only significant limitation to using deconvolution on your images is the time that is required to generate a *z* stack of data that is suitable for deconvolution. Some deconvolution packages will deconvolve a single image, but most require a *z* stack taken with slice intervals significantly better than the recommended optimal sectioning (see section 2.7 for details). As you are trying to put back the high-frequency components into an image, you have to take samples to cover those high frequencies to prevent aliasing. In practice, this means taking *z* sections up to every 0.1 μm under a high-power lens, which can be a very time-consuming process. Therefore, although all confocal images should be deconvolved, this is only necessary if the high-resolution information produces data pertinent to the experimental question.

Acknowledgement

Many thanks to Colin Park from Leica Microsystems (confocal division) for constructive criticism of this manuscript.

3. REFERENCES

1. **Minsky M** (1988) *Scanning*, **10**, 128–138.

★★ 2. **Hibbs AR** (2000) *Confocal Microscopy for Biologists: an Intensive Introductory Course*, 3rd edn. BIOCON, Specialists in Confocal Microscopy, Victoria, Australia. – *An excellent review which provides additional information for the budding biosciences confocal expert.*

3. **Lacaille VG & Androlewicz MJ** (2000) *Traffic*, 1, 884–891.

4. **Cheville NF, Hostetter J, Thomsen BV, Simutis F, Vanloubbeeck Y & Steadham E** (2001) *Dtsch. Tierarztl. Wochenschr.* **108**, 236–243.

5. **Schildgen O, Roggendorf M & Lu M** (2004) *J. Gen. Virol.* **85**, 787–793.

6. **Chong BF, Murphy JE, Kupper TS & Fuhlbrigge RC** (2004) *J. Immunol.* **172**, 1575–1581.

7. **Xu H, Forrester JV, Liversidge J & Crane IJ** (2003) *Invest. Ophthalmol. Vis. Sci.* **44**, 226–234.

8. **Xu H, Manivannan, A, Goatman KA, *et al.*** (2004) *J. Leukoc. Biol.* **75**, 224–232.

9. **Haynes AP, Daniels I, Abhulayha AM, *et al.*** (2002) *Br. J. Haematol.* **118**, 488–494.

10. **Omelyanenko V, Gentry C, Kopeckova P & Kopecek J** (1998) *Int. J. Cancer* **75**, 600–608.

11. **Lee FT, Rigopoulos A, Hall C, *et al.*** (2001) *Cancer Res.* **61**, 4474–8882.

CHAPTER 7

Ultrastructural immunochemistry

Jeremy N. Skepper and Janet M. Powell

1. INTRODUCTION

The landmark publication by Coons *et al.* in 1941 (1) demonstrated that an antibody conjugated to a fluorochrome retained its ability to recognize and bind tightly to its antigen. This was arguably the key event in the development of immunofluorescence microscopy. Once the general principles for immunochemical staining were established, the introduction of particulate markers for transmission electron microscopy (TEM) followed quickly. Ferritin, a moderately electron-dense haem protein, was the first to be conjugated to antibodies some 18 years later (2, 3).

The use of colloidal gold technology was undoubtedly the most significant event in the development of immunochemistry. It was demonstrated that protein molecules, including antibodies, could be adsorbed on to the surface of gold particles with little or no loss of their biological activity (4). Gold particles are particularly useful for TEM studies, as they scatter electrons strongly and even small particles are clearly visible under the electron microscope. Next, the ability to produce colloidal gold particles with different mean sizes and nonoverlapping size-frequency distributions (5) brought the potential for immunochemical staining of multiple antigens on the same thin section, a development that significantly improved the value of the method. The application of random sampling strategies and the use of unbiased stereology allowed us to make quantitative comparisons of labeling density (6, 7). In some instances, it is possible to estimate the concentration of the antigen within its host tissue by making a comparison of labeling density on the specimen with that over an internal standard containing a known concentration of antigen (8).

Before proceeding to immunogold staining, it is important to gather as much information as possible about your antibody and its respective antigen. Where is it likely to be located? Is the antigen extracellular,

Immunohistochemistry: *Methods Express* (S. Renshaw, ed.)

intracellular, membrane-associated, or a soluble component of the cytoplasm? Is it there in significant quantities? Is it sequestered at high concentration in any specific subcellular compartment, such as the mitochondria or the nucleus? How vulnerable to fixation and embedding is the antigen of interest? Information on the specificity of antibodies from Western blotting is valuable, but just because antibodies work well in Western blotting is no guarantee that they will be useful for immunochemistry. Antibodies that 'work well' on blots frequently have to be used at concentrations of up to three or more orders of magnitude greater for immunofluorescence and even more for immunogold staining studies. Some antibodies simply cannot be used for immunochemistry! The degree of resistance of the antigen to fixation is a key issue. In general, the stronger the fixative that can be used, the better the ultrastructure of the thin sections. Unfortunately, the opposite generally applies to the ability of the antibody to bind its antigen. This also relates to cryotechniques. It is possible to freeze and embed tissue at low temperature without any chemical fixation, but ultrastructure is always compromised. If a chemical fixative is added to the substitution mixture, the frozen tissue is dehydrated and fixed at the same time. Fixation is less efficient at low temperature, but as a rule of thumb, the stronger the chemical fixation, the better the ultrastructural preservation and in particular that of membranes. The gold standard is to find the appropriate compromise that allows one to answer the biological question.

For the purpose of this chapter, the terms 'immunochemical' and 'immunogold' can be considered synonymous.

1.1 Fixation and its effect on antigen–antibody binding

Glutaraldehyde and formaldehyde are the two fixatives in most common use, either individually or in combination. Formaldehyde is a monoaldehyde that interacts principally with proteins forming methylene bridges or polyoxymethylene bridges in a concentration-dependent manner. Glutaraldehyde is a dialdehyde that gives superior ultrastructural preservation but causes significant conformational changes to the tertiary structure of proteins. This frequently compromises the ability of an antibody to bind to its antigen. For a detailed discussion on the chemistry of fixation, see (6), (9) and (10), and Chapter 4, section 2.1. In this context, a 'stronger' fixative will be regarded as a fixative containing higher concentrations of the reactive aldehydes.

The ability of the antibody to bind its antigen may be lost at several key stages of processing for TEM: during chemical fixation, dehydration in organic solvents, infiltration with epoxy or acrylic resin, or heat curing or polymerization of the resin. New antibodies should always be tested by a method that does not amplify signal, such as a species-specific, fluorescent secondary antibody method, before proceeding to electron microscopy. There are several key questions to ask if an antibody has been used for prior immunochemical studies:

1. Does the antibody work only on unfixed or cold acetone/alcohol-fixed cryostat sections or cell cultures? If the answer is yes, this antibody may only be usable in methods employing cryoimmobilization and freeze-substitution in pure organic solvent rather than after chemical fixation.
2. At what strength and duration of fixation will the antigens survive and still offer immunogenicity to bind to their respective antibodies?
3. Does the antibody work on sections of formalin-fixed, paraffin wax-embedded tissue, without antigen-retrieval treatment?

It is wise to undertake a systematic evaluation of fixation on a tissue known to contain significant amounts of the antigen under study. This constitutes a positive control, which is highly desirable, if not essential, in any rigorous study and may well provide critical information. Fixation of tissues and organs is best carried out by vascular perfusion (9, 10) to minimize the diffusion distance into the tissue for the fixative. There are, however, circumstances where fixation by perfusion is impossible or may be undesirable. Bendayan *et al.* (11) showed that immunogold staining of serum albumin in glomerular capillaries was reduced dramatically after perfusion fixation, presumably because the serum albumin molecules were washed out during exsanguination.

If it is not possible to fix by perfusion, for example when working with human tissues from surgical material or biopsies, samples should be small. A simple method of achieving uniformity of fixation is to glue two safety razor blades together at the shank, to produce two parallel blades <1 mm apart. Tissues are sampled using a gentle slicing motion, rather than by applying significant vertical force, in order to minimize mechanical damage. Alternatively, a tissue chopper or vibrating microtome can be used to cut thin slices. Cells in culture are much easier to deal with, as diffusion distances for fixatives are minimal. They should be cooled to 4°C and rinsed in normal saline (0.9%, w/v, sodium chloride) before fixation. Nonadherent cells can be fixed in suspension, whilst adherent cells should be fixed *in situ* for 30–60 min and then scraped free of their substrate.

The temperature and duration of fixation should both be standardized to maintain uniformity between experiments. We carry out initial fixation tests at 4°C for no more than 120 min for tissues and

30–60 min for cell cultures, whilst others prefer fixation at 37°C (12), but only trial and error will determine the appropriate compromise between structural preservation and the ability of the fixed antigen to bind antibody. Safety is a major issue when fixation is carried out at temperatures above 4°C as aldehydes are volatile, formaldehyde is a known carcinogen and glutaraldehyde can cause occupational asthma. If fixation is performed at an elevated temperature, it should be carried out in a fume hood.

It is convenient to test new antibodies on adherent cell cultures expressing the antigen of interest, grown on glass coverslips, or on cryostat sections. Cells are grown to near-confluence on 19 mm diameter coverslips of No. 1 thickness and fixed for 30–60 min at 4°C. Naturally, if the antigens will survive longer periods of fixation (up to 4 h), then ultrastructural preservation will be even better. They are rinsed in four to six changes of buffer before being stained immunochemically. Alternatively, fixed tissues are infused with 20% (w/v) sucrose and frozen to prepare cryostat sections. We routinely store a range of fixed and unfixed tissues (myocardium, liver, gut, placenta, etc.) under liquid nitrogen so that material is always available for testing new antibodies. An initial test is carried out comparing the effects of weak and strong fixatives using the following solutions:

(a) 1% (w/v) formaldehyde in 0.1 M PIPES or HEPES buffer (pH 7.4) containing 3 mmol/l calcium chloride.
(b) 4% (w/v) formaldehyde in 0.1 M PIPES or HEPES buffer (pH 7.4) containing 3 mmol/l calcium chloride.
(c) 8% (w/v) formaldehyde in 0.1 M PIPES or HEPES buffer (pH 7.4) containing 3 mmol/l calcium chloride.
(d) 3% (w/v) formaldehyde plus 0.05–0.5% (w/v) glutaraldehyde in 0.1 M PIPES or HEPES buffer (pH 7.4) containing 3 mmol/l calcium chloride.

The fixatives shown above are listed in ascending order of potential ultrastructural preservation but probable descending order of antibody binding. Tissue sections or cultured cells fixed in solutions (a), (b), and (c) are ready for immunochemical staining after rinsing in buffer. Those fixed in solution (d) must be incubated in 0.5% (w/v) sodium borohydride for 5–10 min and rinsed in buffer to quench the autofluorescence generated by glutaraldehyde. Test parameters should also include a range of dilutions of primary antibodies, usually 1:5, 1:25, 1:100, and 1:1000 for monoclonal antibodies and 1:50, 1:250, 1:1000, and 1:5000 for polyclonal antibodies.

Antigens that survive very strong fixation and embedding in paraffin wax may well survive ambient temperature dehydration and embedding in thermally cured or polymerized epoxy resin after secondary fixation with osmium tetroxide. In this method, the osmium tetroxide is removed from the superficial layers of the section by treatment with periodic acid and/or sodium metaperiodate (13). Thin sections are floated on drops of the oxidizing agent of choice, e.g. 5% (w/v) sodium metaperiodate, for 10–20 min, and then rinsed thoroughly with ultrapure water before commencing immunochemical staining. Periodic acid and sodium metaperiodate are both oxidizing agents with differing efficacies. Some workers just use sodium metaperiodate, whilst others suggest that a sequential treatment with both produces stronger immunochemical staining. We tend to use a single treatment with sodium metaperiodate, which removes osmium tetroxide from the surface of the thin section, and in some cases this will enhance the binding of an antibody to its antigen at that surface. Antigens that withstand modest fixation but not paraffin wax embedding are generally more suitable for embedding in acrylic resin at ambient or at low temperature. Antibodies that only work on unfixed cryostat sections may work in cells or tissues that have been cryoimmobilized, dehydrated by freeze-substitution, and embedded at low temperature. However, there is no guarantee that the integrity of the antigen will not be compromised by the subsequent dehydration, embedding, and curing or polymerization of the resin.

It may also be necessary to use a stronger fixative to retain antigens that are freely soluble in the cytoplasm (14). It is interesting to note that cells with a high content of secretory granules and endoplasmic reticulum often show reasonable preservation, even after weak fixation, particularly if they are processed subsequently using the freeze-substitution and low-temperature embedding route. This may be at least partly due to their high protein content (see *Fig. 1*). As the strength of fixation is reduced, ultrastructural preservation becomes poorer, particularly that of membranes. The low-temperature methods compensate to some degree, but it is inevitable that weaker fixation means poorer preservation. The method that retains the best membrane preservation is undoubtedly the ultrathin, thawed cryosection or 'Tokuyasu' method (15), but again stronger fixation gives better preservation. A comprehensive description of this technique is beyond the scope of this chapter and the reader is referred to (6) and the seminal papers by Peters *et al.* and Liou *et al.* (12, 16).

1.2 Controls

Controls are essential but ostensibly simple, requiring tissue or cells expressing significant amounts of the antigen as a positive control. A

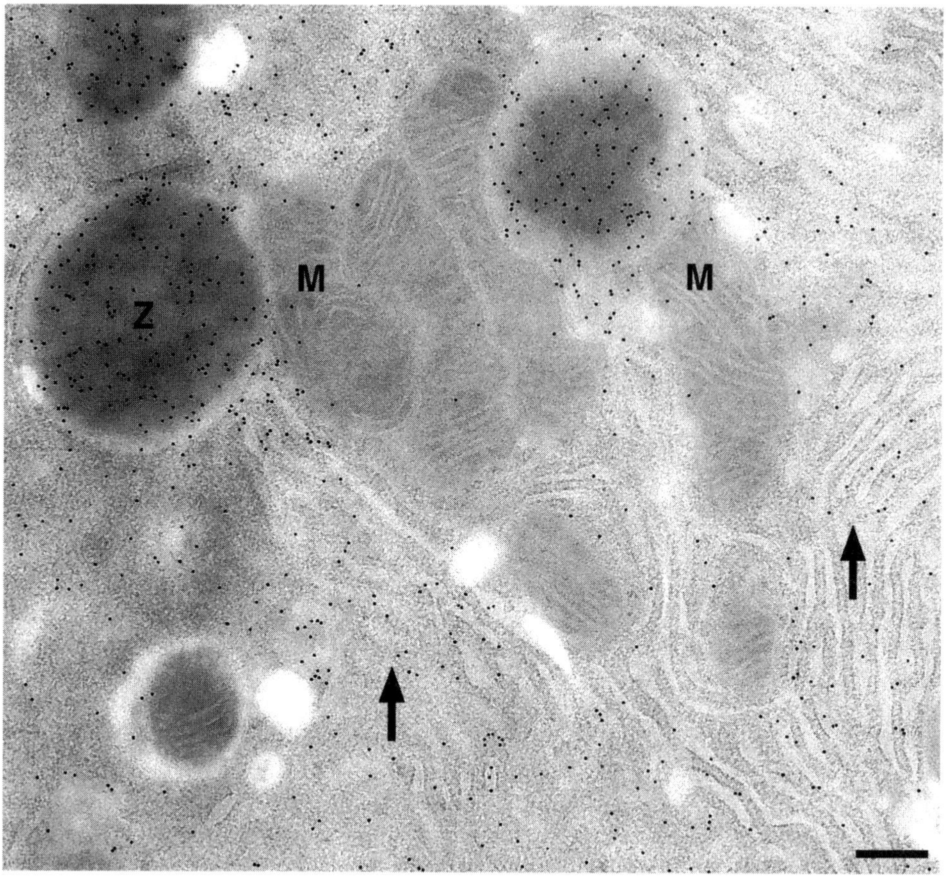

Figure 1. Thin section of a rat pancreatic acinar cell.
The section was fixed in 3% formaldehyde, cryoprotected in 30% polypropylene glycol, dehydrated by freeze substitution, and low-temperature embedded in Lowicryl HM20. Cells were immunolabeled for the presence of amylase. Gold particles indicate the rough endoplasmic reticulum (arrows) and zymogen granules (Z). Mitochondria (M) are unlabeled, showing that nonspecific labeling is low. Bar, 200 nm.

negative control is equally important, as it will indicate whether there is nonspecific binding of primary or secondary antibodies. Sections of cells should also be exposed routinely to the secondary antibody alone to be certain that there is no nonspecific binding to any component of the tissue. In a recent unpublished study carried out with Raghu Padinjat of the Babraham Institute (Cambridge, UK), we encountered a most elegant example of a combined positive and negative control in adjacent cells of the same tissue. Omatidia are the light-sensing structures of the *Drosophila* eye. Each omatidia contains seven rhabdomeres (see *Fig. 2*), which have extensive membrane systems derived from microvilli. The membranes of six of the rhabdomeres are rich in rhodopsin, a light-absorbing pigment, whilst the seventh rhabdomere contains no rhodopsin (see *Fig. 2*), making it an ideal negative control. If excessive nonspecific binding of the primary or secondary antibody is apparent, protein can be added to the buffers to inhibit it competitively. Various proteins are used for this purpose, but in our hands BSA or coldwater-fish gelatin, both used at 0.5–4% (w/v), give the most consistent results. Remember this will also

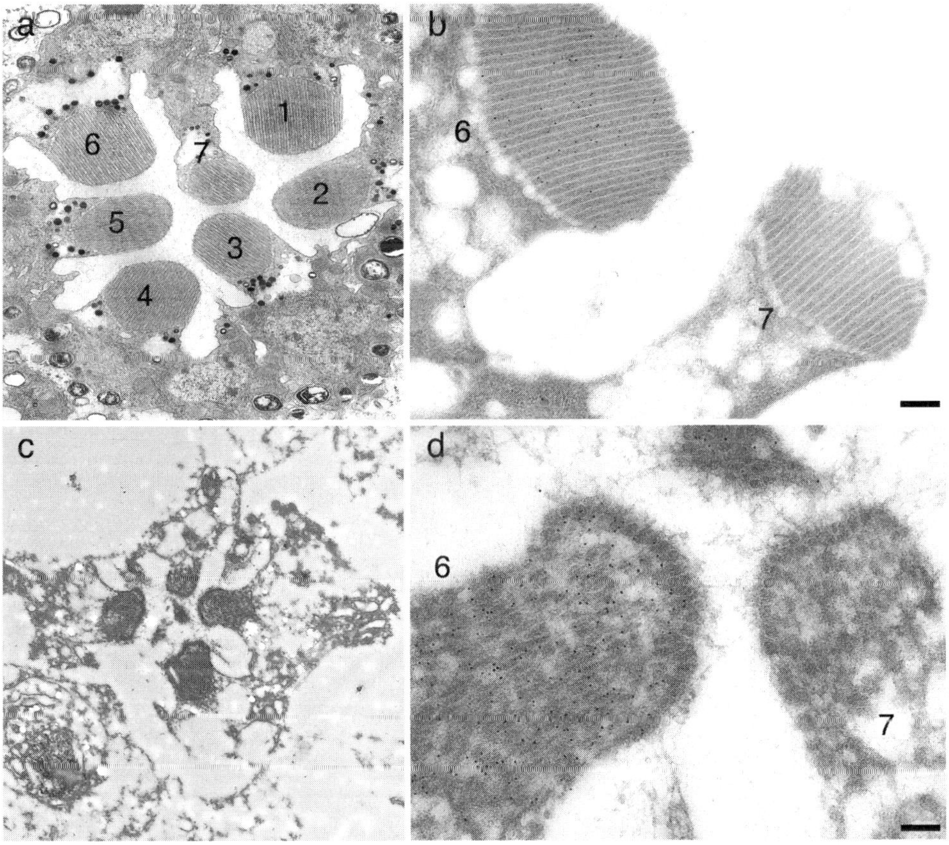

Figure 2. Thin sections through a single omatidia from a wild-type or mutant *Drosophila* eye, immunolabeled for rhodopsin.
The eyes were fixed in 3% glutaraldehyde, osmicated, and embedded in Spurr's resin (*a*, *b*; wild type) or fixed in 4% formaldehyde and embedded in LR White (*c*, *d*; mutant). Each omatidia contains seven rhabdomeres (*a*). Rhabdomeres 1–6 express rhodopsin, whilst rhabdomere 7 does not (*b*). In the mutant eye, omatidia are deleted or altered (*c*). The rhabdomeres are also structurally altered, but their staining pattern for rhodopsin remains unchanged, with no expression of rhodopsin in rhabdomere 7 (*d*). Bars, 200 nm.

competitively inhibit specific binding, so the concentration of blocking protein should be kept as low as possible.

1.3 Why do we need to use electron microscopy?

The answer to this is resolution. In the light, confocal, and two-photon microscopes, resolution is diffraction-limited to 180–200 nm in the *x-y* axes, dependant on the numerical aperture of the objective lens and the wavelength of light used to generate the image. Resolution in the *z* axis is much poorer at 500–600 nm. However, there are techniques that can bypass these limitations. These include total internal reflectance (TIRF) microscopy, stimulated emission depletion microscopy and 4Pi microscopy. TIRF (17) and 4Pi (18) microscopy can exceed 100 nm resolution, but with severe limitations on specimen and lens geometry in 4Pi microscopy and in the depth of imaging into a sample with TIRF microscopy. Stimulated emission depletion microscopy (19) can exceed 50 nm resolution, but requires a very high signal-to-noise ratio and an almost ideal sample.

1.4 Quantification

If a single compartment is being stained immunochemically and the biological question is simply whether or not there is staining over that compartment, then quantification is unnecessary. If label (staining) density is low and you wish to make a comparison between multiple compartments in control and experimental subjects, then quantification is essential. Quantification of label density is simple and strengthens data immensely. It is a simple extension of stereology, which is used to gain three-dimensional data from (effectively) two-dimensional sections. When comparing mutant and wild-type organs, it is desirable to start the comparison with an estimate of the volume or reference space of the organ itself. If the organ of the mutant is halved in volume but the percentage of it occupied by a specific cell type is doubled, the total volume of that cell type is unchanged. This phenomenon is known as the 'reference trap' (20). A typical example might be to examine the effect of a mutation on the distribution of rhodopsin in the eye of *Drosophila* (see *Fig. 2*). After fixation and embedding in a suitable resin, serial sections (2 μm in thickness) are cut through the eye and the Cavalieri method (21) is used to estimate the volume of the eye in mutant and wild-type flies. At four randomly selected levels through the layer containing the rhabdomeres, thin (50–70 nm in thickness) sections are cut, immunogold labeled for rhodopsin and contrast counterstained with uranyl acetate and lead citrate. Both uranyl acetate and lead citrate impart contrast to the tissue. They are viewed at 80 kV in a transmission electron microscope using a 10 or 20 μm objective aperture to maximize contrast. A quadratic (square) lattice is overlaid on the TEM image and the volume fraction (Vv, expressed as a percentage) of the eye occupied by omatidia and rhabdomeres is estimated by point counting (21), i.e. counting the number of points (P) from the intersections of the counting lattice that overlie the area of interest (i). The formula for this calculation is:

$$Vv_{rhabdomere}\ (\%) = (Pi_{rhabdomere}/Pi_{total}) \times 100$$

where $Vv_{rhabdomere}$ is the percentage of the eye occupied by rhabdomeres, $Pi_{rhabdomere}$ is the number of lattice intersections overlying rhabdomeres and Pi_{total} is the total number of lattice intersections overlying all compartments of the eye. Therefore, if $Pi_{rhabdomere} = 10$ and $Pi_{total} = 100$, 10% of the eye is occupied by rhabdomeres.

The light-absorbing pigment rhodopsin is associated with the photoreceptor membranes of the rhabdomere and immunogold label density can be calculated as the number of gold particles per unit area (number/square micrometer) of rhabdomere. Number/unit area can be estimated by randomly selecting squares of the counting lattice overlying the areas of interest (rhabdomeres) and counting all of the gold particles within the square and those within the frame that also intersect with two of the four boundary lines of the counting frame. This is known as the forbidden line rule (22) and prevents the underestimation of particle density that occurs if particles are only counted if they are within the square but not touching the counting frame. Similarly, the number will be overestimated if all particles, including those intersecting all four boundaries, are included. The area of an individual square counting frame of a quadratic lattice is D^2, where D is the distance between two intersections of the lattice. Therefore, gold label density can be estimated by summing the number of gold particles in, for example, ten randomly selected test frames and dividing that by the total area of those frames in micrometers.

The procedure described above gives a parametric estimate of gold labeling density over a structure or series of subcellular compartments. Mayhew and co-workers (23, 24) have suggested a nonparametric alternative that estimates the 'relative labeling density' between compartments and between control and experimental subjects.

2. METHODS AND APPROACHES

As one would expect with a technology that is more than 40 years old, the number of methods and their variants is extensive. Many methods are incompletely described, particularly those in research publications with restrictions on space for methods, and to the novice many may appear to be a combination of 'cookery' and 'witchcraft'. This chapter will describe four methods that are in common use. For more comprehensive descriptions of the range of techniques available, see (6) and (25).

Three of the methods described here are post-embedding methods. In these methods, the cells or tissues are fixed chemically or cryoimmobilized, dehydrated, and embedded in epoxy or acrylic resins. Thin sections (50–70 nm in thickness) are cut using an ultramicrotome with a diamond knife, using a water bath to collect the sections as they slide off the knife. The sections are stretched with solvent vapor or a heat source and collected on to either bare or plastic-coated nickel grids. The sections are then stained immunochemically with primary antibodies raised against antigens exposed on the surface of the sections. The primary antibodies are then visualized by staining immunochemically with secondary antibodies raised against the species and isotype of the primary antibodies, conjugated to colloidal gold particles. The immunochemically stained sections are then contrast stained with salts of uranium (uranyl acetate) and lead (lead citrate) to reveal the ultrastructure of the cells, and are finally viewed by TEM.

The fourth method is a pre-embedding method that is used if antigens are damaged by resin embedding, or if the best preservation of membranes is required. Cells or tissues are fixed as strongly as possible and then treated with a cryoprotectant, which is usually a mixture of sucrose and polyvinylpyrrolidone. They are frozen on to pins in liquid nitrogen and sectioned at −100°C. The frozen sections are thaw-mounted on to Formvar/nickel film grids and the cryoprotectant is removed by floating the grids on drops of phosphate-buffered saline. The immunogold staining is performed on the unembedded section and it is subsequently contrast counterstained and infiltrated with a mixture of methylcellulose and uranyl acetate. Methylcellulose reduces shrinkage of the section when it is air dried before viewing by TEM.

2.1 Epoxy resin sections

Chemical fixation and embedding in a highly cross-linked epoxy resin is the method of choice for optimal ultrastructure and stability of the thin section in the electron beam. Paraffin wax cannot be used for TEM as it is impossible to cut thin enough sections, since the wax is too soft. Even if it were possible to cut sections that were thin enough, the wax would evaporate in the electron beam and contaminate the column of the microscope. Immunogold staining of thin epoxy resin sections is useful if the antigen of interest is very resistant to fixative or if only archived material that was fixed primarily for ultrastructural studies is available. It would be ideal if we could fix and embed tissue to produce the very best ultrastructure, yet leave the tissue with sufficient antigenicity for it to be immunochemically stained. This would optimally include fixation in a high concentration of glutaraldehyde (2.5%, w/v, or higher, see Chapter 4, section 2.1.1), followed by secondary fixation with osmium tetroxide and bulk staining in uranyl acetate. Osmium tetroxide fixes by binding to double bonds in unsaturated fatty acids, retaining them in the subsequent dehydration in organic solvent. It adds positive contrast, since it is a heavy metal that scatters electrons. Similarly, uranyl acetate acts as both a fixative and a stain, as it helps retain phospholipids and adds contrast to the thin sections by scattering electrons. The fixed tissue is dehydrated in an organic solvent infiltrated with an epoxy resin, which is thermally cured at 60°C for up to 48 h. Epoxy resin monomers are joined end to end to form long-chain polymers, which are in turn cross-linked to adjacent polymers during the curing process. This makes them very stable in the transmission electron microscope but hinders access of the antibody to the antigen. Some antigens do survive this treatment, notably small peptide hormones or neurotransmitter substances that are found highly concentrated in secretory vesicles (see *Fig. 3*). High concentrations of glutaraldehyde are used in protocols for immunochemical staining of amino acid neurotransmitters, such as glutamate and γ-aminobutyric acid (8). This appears to be necessary to ensure they are not physically extracted during subsequent dehydration and embedding. It is generally necessary to remove osmium tetroxide from the superficial regions of the thin section to be immunochemically stained. This is readily achieved by treatment with one or a combination of the following oxidizing agents: 10% (v/v) hydrogen peroxide (26), 4% (w/v) sodium metaperiodate (27), or 1% (w/v) periodic acid (8). This pre-treatment of resin sections of tissues fixed to maximize ultrastructural preservation has been used to great effect for the study of secretory proteins, peptides, and neurotransmitters (13, 28–31). However, these antigens are generally present in very high local concentrations within secretory granules. Oxidizing agents attack the hydrophobic alkane side-chains of epoxy resins, which makes the sections more hydrophilic (26). This allows more intimate contact between the immunochemical reagents and the antigens exposed at the surface of the sections.

2.2 The acrylic resins London Resin (LR) White and Gold

LR White was introduced as a low-toxicity alternative to epoxy resins, which frequently contained carcinogens (32). It contains an initiator and can be polymerized by the application of heat at 48–50°C or by chemical catalysis at temperatures as low as −15 to −20°C, albeit exothermically. At

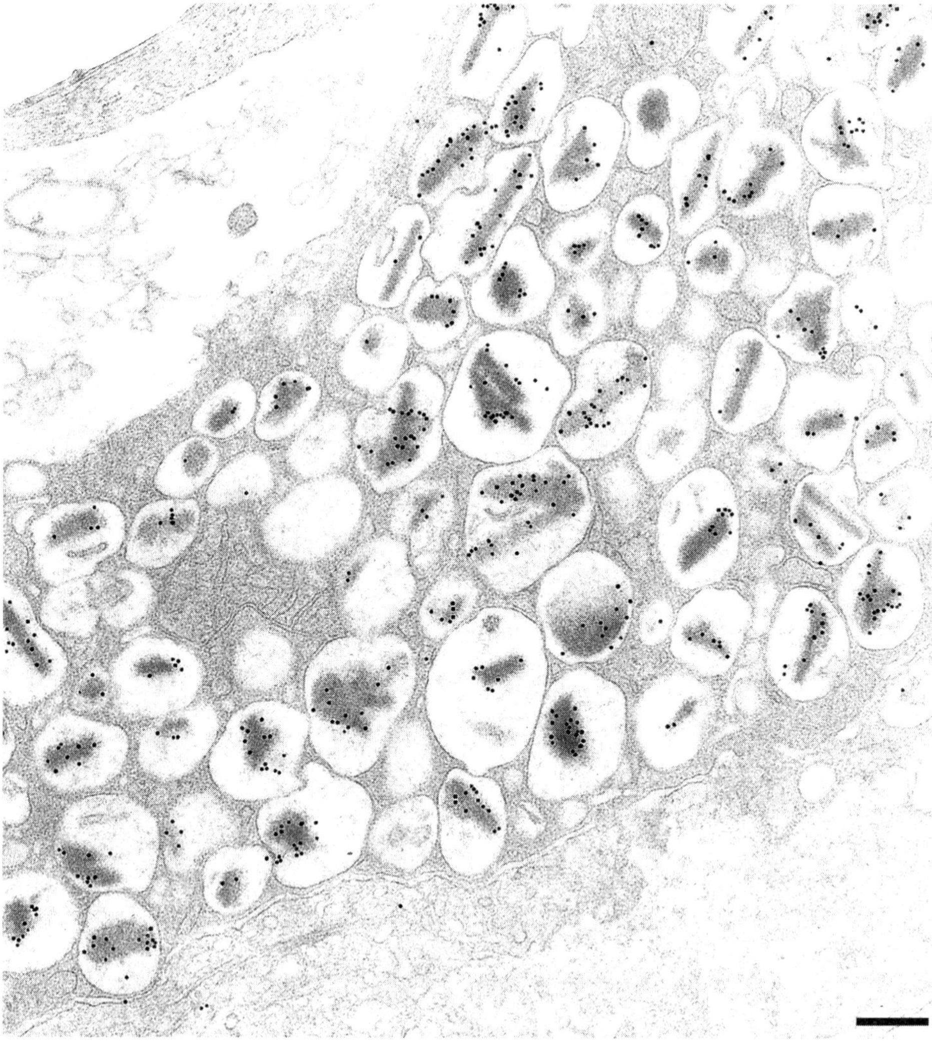

Figure 3. Thin section through a rat pancreatic β-cell.
The tissue was fixed in 4% glutaraldehyde/1% osmium tetroxide, bulk stained in uranyl acetate, and embedded in Spurr's resin. The section was treated with sodium metaperiodate before immunolabeling for insulin. The crystalline cores of the secretory granules are heavily labeled with gold particles. Bar, 250 nm.

temperatures below −15°C, its viscosity is very high and infiltration of the resin into the tissue becomes problematic. LR Gold is less hydrophobic and can be polymerized by photo-initiation using benzoin methyl ether as a catalyst down to −25°C. It should be noted that at temperatures below −18°C, the initiator can spontaneously come out of solution. Unlike the simplest acrylic resins, in which monomers are polymerized to form long chains, the LR resins contain aromatic cross-linkers to improve the stability of the sections under the electron beam. LR White and Gold both have very low viscosity and readily penetrate, even into dense tissue. Aldehyde-fixed tissue is dehydrated and embedded in the acrylic resin without secondary fixation in osmium tetroxide. The tissue is dehydrated in ethanol, impregnated in acrylic resin, and polymerized under vacuum or in a nitrogen atmosphere. The inert atmosphere is necessary because oxygen inhibits the polymerization of these resins. Acetone is not recommended as a dehydrating agent, as it can act as a scavenger of free radicals, which can interfere with the polymerization of the resin. A convenient method for flat embedding is to place tissue in an aluminum weighing boat and exclude oxygen by dropping a piece of Melinex sheet or a Thermanox coverslip on to a positive meniscus of resin. Polymerization can be initiated chemically or photolytically at 4–20°C or thermally at 48–60°C. It is claimed that reducing the temperature during polymerization enhances antigen survival. If this is the case, in most instances the gain is likely to be marginal if the difference is a drop from 60°C to ambient temperature. Membrane preservation can be improved by bulk staining with uranyl acetate (see *Fig. 4*), before

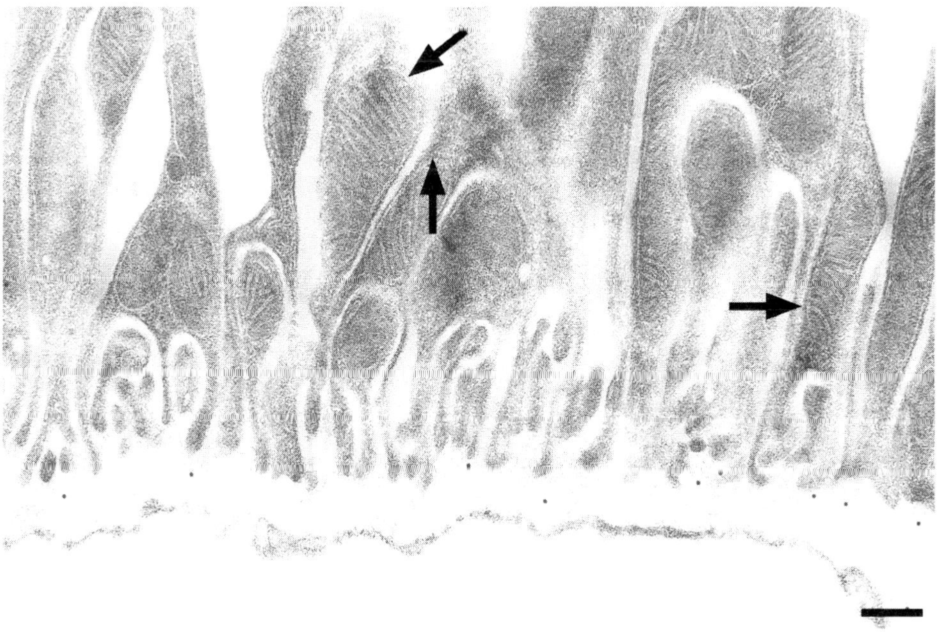

Figure 4. Thin section through a proximal convoluted tubule of a rat kidney.
The tissue was fixed by immersion in 2% formaldehyde and embedded in LR White after bulk staining in uranyl acetate. The basal lamina is labeled with gold particles after immunostaining for laminin. Despite the weak fixation, the outer mitochondrial membranes and cristae of mitochondria (arrows) can be clearly distinguished. Bar, 250 nm.

dehydration (33). The advantages of this method are: (i) the polymerized acrylic resin matrix is 'looser' than that produced in a cured epoxy resin and (ii) the sectioning properties are different to those of epoxy resins and the antigens revealed at the surface of the section may be more accessible to the antibody molecules.

2.3 Freeze substitution and low-temperature embedding in Lowicryl HM20

Lightly fixed pieces of tissue are cryoprotected by immersion in 30% (v/v) glycerol or polypropylene glycol (34). The cryoprotectant provides many nucleation sites within the tissue, so that, even when slower freezing methods are used, the small ice crystals formed are unresolvable by TEM at the magnifications used for most immunogold staining studies. The cryoprotected tissues are mounted on small pieces of aluminum foil or on pieces of Millipore filter. They are quench frozen, by plunging into liquid propane cooled in liquid nitrogen. Adequate freezing can also be obtained using nitrogen slush or even liquid nitrogen for very small samples. Alternatively, monolayers of cells or thin slices can be frozen rapidly, freeze substituted and low-temperature embedded with no chemical fixation at all (35, 36) (see *Fig. 5*). This method is therefore suited to antigens that are sensitive to aldehyde fixation.

The frozen tissue samples are transferred under liquid nitrogen to vials half filled with frozen methanol or methanol containing low concentrations (0.01–1%, w/v) of uranyl acetate. Chilled metal forceps are used to move samples and the tissues frequently develop a 'charge', causing them to stick to the forceps. A cooled wooden cocktail stick can be used to dislodge them, or ceramic forceps can be used if necessary. The vials of frozen tissue and substitution material are transferred to a substitution vessel where the temperature can be controlled and a nitrogen atmosphere maintained.

Once the samples are in the substitution vessel, the temperature is raised typically at 5°C/h to −90°C and this temperature is maintained for 24 h. This temperature is cold enough to prevent recrystallization of water and thus tissue disruption, but high enough for movement of water to take place and allow substitution with the liquid methanol. After 24 h, approximately 90% of the water has been substituted. The substitution medium is replaced and the temperature is raised to −70°C for 24 h. The substitution medium is changed again and the temperature is raised to −50°C. The tissue is impregnated with Lowicryl HM20 over a period of 1–5 days and the resin is polymerized by UV irradiation at −50°C. This method has been used successfully to localize adhesion molecules (34), which are

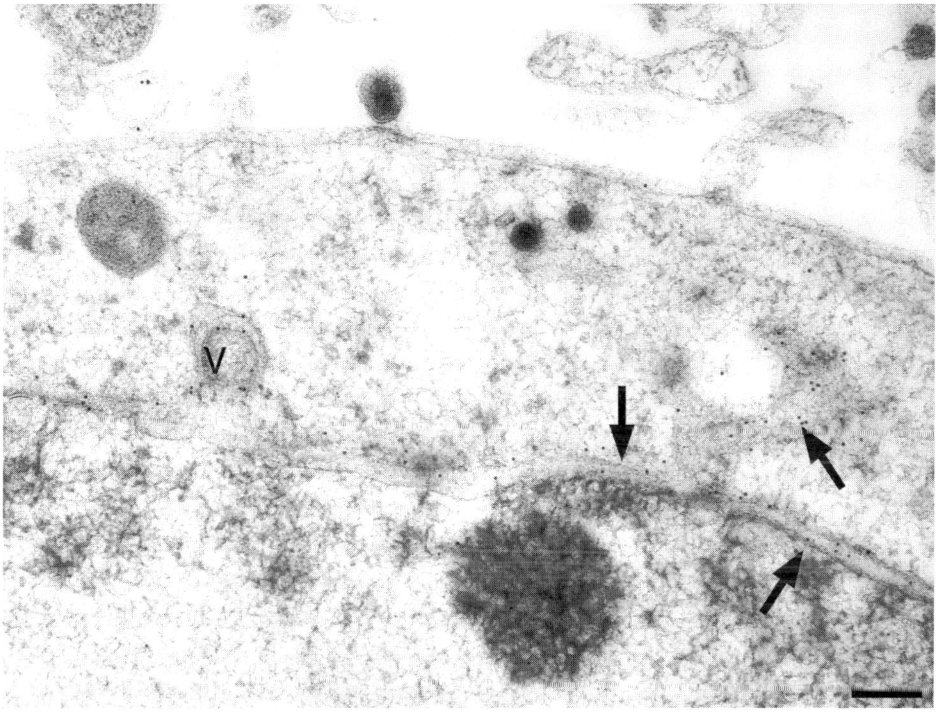

Figure 5. Thin section of a Vero cell infected with human papilloma virus.
Cells were quench frozen in melting propane cooled in liquid nitrogen, dehydrated by freeze substitution against pure methanol containing 0.1% uranyl acetate, and low-temperature embedded in Lowicryl HM20. Cells were immunolabeled for glycoprotein D. Gold particles indicate the nuclear membrane and the rough endoplasmic reticulum (arrows) and the membrane acquired by a virus particle (V) that has just budded through the nuclear envelope. Bar, 200 nm.

notoriously labile during fixation and embedding. This is the simplest and most versatile of the post-embedding procedures.

2.4 Ultrathin thawed cryosections

In this technique, the sample is sectioned at low temperature, thaw-mounted on to film grids, immunochemically stained, contrast counterstained, and embedded/encapsulated *in situ* on the grid. Applying immunogold reagents to sections of lightly fixed tissue, free of embedding medium, can be a very sensitive method of immunochemical staining. This technique is frequently referred to as the Tokuyasu technique (37–40) after its pioneer. It is one of the few methods that is consistently used for immunochemical staining of sparse and labile membrane-bound proteins such as receptor molecules. It has been used to great advantage in the study of receptor internalization and endocytosis (41, 42).

Small cubes of fixed tissue (<0.25 mm^3) impregnated with either 2.3 M sucrose (39) or a mixture of 1.9–2.1 M sucrose and 10% (w/v) polyvinylpyrrolidone (40) (to act as cryoprotectants) are mounted on pins and frozen in liquid nitrogen. The high concentrations of cryoprotectant make rapid freezing unnecessary. The frozen samples are transferred to an ultramicrotome with a cryochamber and sectioned at a temperature of between -90 and -130°C. In tissues with voids such as blood vessels, the sucrose tends to crumble rather than section. This effect is particularly noticeable in fragile embryos. It can be prevented by filling the lumen of blood vessels with gelatin (43) or by infiltration of embryos with polyacrylamide gel (38).

Thin sections are maneuvered into position away from the cutting edge of the knife with an eyelash and retrieved on a drop of cold 2.3 M sucrose, or sucrose and polyvinylpyrrolidone, held in a copper loop with an internal diameter of 1–1.5 mm. The droplet of sucrose is moved rapidly towards the sections, which will jump towards it and 'disappear'. The sucrose must remain liquid while the sections are retrieved or they will not fully decompress, so speed is fundamental. Recently introduced alternative retrieval fluids are a 50:50 mixture of 2% (w/v) methylcellulose and 2.3 M sucrose or 1.5–2% (w/v) methyl cellulose, and 0.3–3% (w/v) uranyl acetate. If it is possible to use strong fixation (4–8%, w/v, formaldehyde for 2–4 h), the preservation of cell membranes is excellent (16) (see *Fig. 6*). These sections can be stored on buffer at 4°C for several hours or even overnight, if more blocks are to be sectioned. However, caution should be exercised if the tissue has been fixed very lightly, as the ultrastructure will deteriorate as a function of the time that the section is floated on the buffer.

Immunogold staining is carried out essentially the same as for resin sections with a few modifications. As the tissue has been fixed and sectioned directly after cryoprotection, residual reactive aldehyde groups may remain in the sections. These are quenched by exposure to 0.1–1% (w/v) lysine or glycine in phosphate- or Tris-buffered saline for 10 min. The absence of an embedding medium means that many primary antibodies will bind strongly after 0.5–2 h exposure, so it is often convenient to use more-dilute primary antibody solutions and stain immunochemically overnight. This conserves antibodies and tends to produce less nonspecific background staining.

After immunochemical staining, the sections are contrast counterstained and encapsulated in a matrix to prevent gross collapse of the section caused by surface tension effects during subsequent air drying.

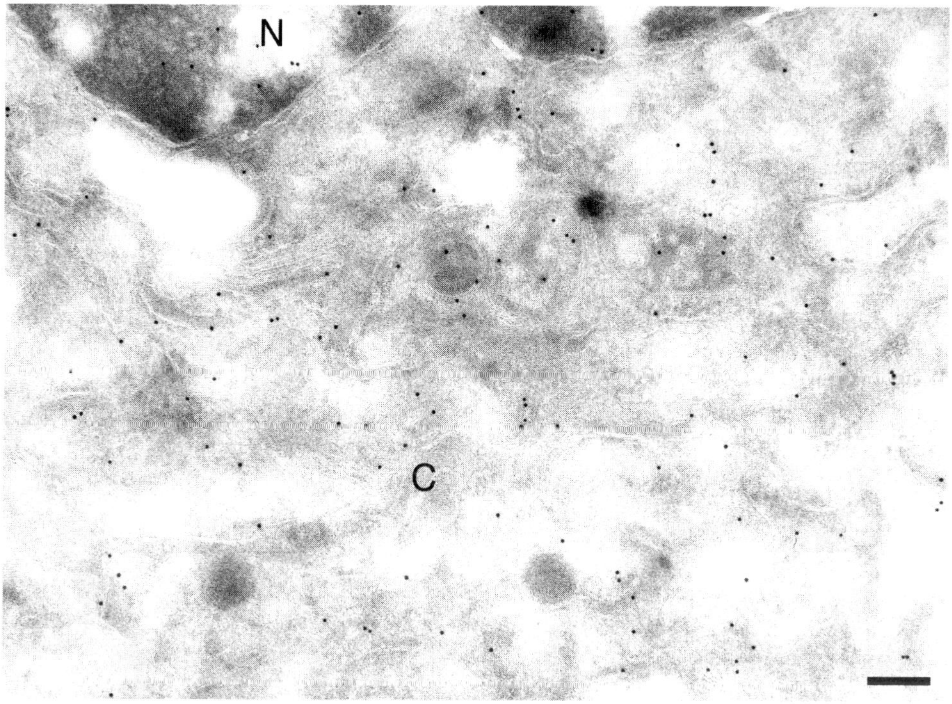

Figure 6. Ultrathin thawed cryosection of placental syncytium.
Sections were fixed in 6% formaldehyde, cryoprotected in sucrose and polyvinylpyrrolidone, and retrieved from the microtome on methylcellulose and sucrose. Cells were immunolabeled for copper/zinc superoxide dismutase. Gold particles can be seen over both the cytoplasm (C) and the nucleoplasm (N). Bar, 200 nm.

The most commonly used stain is uranyl acetate, which produces a negative contrast. Many variations and alternatives to uranyl acetate have been proposed. These have been discussed in detail by Griffiths (6). The simplest method is to rinse sections briefly in cold deionized water and incubate them on drops of 2% (w/v) methylcellulose and 3% (w/v) aqueous uranyl acetate in ratios varying from 9:1 to 5:1. Excess stain is blotted away with hardened filter paper and they are air dried before viewing. Sections prepared in this way are stable for a considerable time. The thickness of the section and that of the final embedding layer influence the contrast in the transmission electron microscope. The thinnest, flattest, sections produced with diamond knives produce the best contrast between the section and antibodies conjugated to small colloidal gold particles (5–20 nm) (12).

2.5 Recommended protocols

Protocol 1

Immunogold staining of epoxy resin sections

Materials and Reagents (please refer to *Appendix 1* for recipes)

- 400 Mesh nickel grids
- 1% (w/v) Coldwater-fish gelatin[a] in PBS containing 0.001% (v/v) Tween 20 and Triton X-100[b] (PBSG)
- Dental wax[c] (or parafilm)
- Diamond trim tool and 45° ultradiamond knife (Diatome AG)
- EM UCT ultramicrotome (Leica Microsystems)
- FEI Tecnai 120 TEM
- Lead citrate
- 50% (v/v) Methanol
- 50% (v/v) Methanol containing saturated uranyl acetate
- Dulbecco's 'A' PBS (pH 7.6)
- 1% (w/v) Aqueous periodic acid[d]
- Primary antibodies optimally diluted in PBSG
- Secondary antibodies, optimally diluted in PBSG and raised against the species of the primary antibody and conjugated to 10 or 15 nm colloidal gold particles
- 4% (w/v) Aqueous sodium metaperiodate[e]
- Ultrapure water

Method

CARRY OUT ALL NICKEL GRID INCUBATIONS/RINSES ON DENTAL WAX

1. Cut thin sections of 50–70 nm and mount on to nickel grids.
2. Incubate sections on drops of 4% (w/v) aqueous sodium metaperiodate for 10 min at room temperature.
3. Rinse grids in ultrapure water for 30–40 s.
4. Incubate sections on drops of 1% (w/v) aqueous periodic acid for 10 min.
5. Rinse in ultrapure water for 30–40 s.
6. Incubate sections on drops of PBSG for 10 min.
7. Incubate sections overnight on drops of PBSG containing optimally diluted primary antibodies.
8. Rinse sections on ten 100 μl drops of PBS for 2 min on each drop.
9. Incubate sections on drops of PBSG containing optimally diluted species-specific secondary antibodies conjugated to 10 or 15 nm gold particles at room temperature for 2 h.
10. Rinse sections in ultrapure water for 30–40 s.
11. Counterstain sections by floating grids section side down on drops of 50% (v/v) methanol containing saturated uranyl acetate for 0.5–10 min at room temperature, followed by a rinse in 50% (v/v) methanol and a rinse in ultrapure water (44).

12. Counterstain sections by floating grids section side down on drops of lead citrate (45) for 0.5–10 min at room temperature in a Petri dish containing a few grains of moistened potassium hydroxide (to prevent lead carbonate precipitation).
13. Rinse grids extensively in ultrapure water and view at 80 kV in a transmission electron microscope.

Notes

[a]Used as a competitive inhibitor of nonspecific staining.
[b]Used as a detergent to facilitate access of antibody to antigen.
[c]Used as a clean hydrophobic surface to perform immunogold staining of thin sections mounted on TEM grids and floated on small drops of reagents.
[d]Used as an oxidizing agent to remove osmium tetroxide from the surface of thin sections. In some cases, this will enhance the binding of an antibody to its antigen at that surface.
[e]Used to remove osmium tetroxide from the surface of the thin section.

Protocol 2

Immunogold staining of LR White resin sections

Please refer to the relevant notes in *Protocol 1*.

Materials and Reagents (please refer to *Appendix 1* for recipes)

- 400 Mesh nickel film grids
- Aluminum weighing boats
- 1% (w/v) Coldwater-fish gelatin[a] in PBS containing 0.001% (v/v) Tween 20 and Triton X-100[b] (PBSG)
- Dental wax[c] (or parafilm)
- Diamond trim tool and 45° ultradiamond knife (Diatome AG)
- EM UCT ultramicrotome (Leica Microsystems)
- 70% (v/v) Ethanol
- 95% (v/v) Ethanol
- 100% Ethanol
- FEI Tecnai 120 transmission electron microscope
- 4% (w/v) Formaldehyde (made from freshly depolymerized paraformaldehyde) in 0.1 M PIPES buffer (pH 7.4) containing 2 mmol/l calcium chloride[f]
- Gelatin capsules
- Lead citrate
- 50:50 Mixture of 100% LR White resin (hard consistency) and 100% ethanol
- Melinex polyester sheet
- 50% (v/v) Methanol
- 50% (v/v) Methanol containing saturated uranyl acetate
- Dulbecco's 'A' PBS (pH 7.6)
- 0.1 M PIPES buffer (pH 7.4)
- Primary antibodies optimally diluted in PBSG
- Secondary antibodies, optimally diluted in PBSG and raised against the species of the primary antibody and conjugated to 10 or 15 nm colloidal gold particles
- 0.9% (w/v) Sodium chloride

- Ultrapure water
- 50% (v/v) Methanol saturated with uranyl acetate

Method

1. Rinse cells or small pieces of tissue twice in 0.9% (w/v) sodium chloride.
2. Incubate in 4% (w/v) formaldehyde (made from freshly depolymerized paraformaldehyde) in 0.1 M PIPES buffer (pH 7.4) containing 2 mmol/l calcium chloride for 1 h at 4°C. If the cells are adherent, scrape them free from the substrate and transfer to 1.5 ml tubes.
3. Rinse cells or small pieces of tissue four times in 0.1 M PIPES buffer over a period of 20 min and twice in ultrapure water.
4. Incubate cells or small pieces of tissue in 2% (w/v) aqueous uranyl acetate for 30 min at room temperature and rinse three times in ultrapure water.
5. Dehydrate cells or small pieces of tissue in three changes of 70% (v/v) ethanol, three changes of 95% (v/v) ethanol and three changes of 100% ethanol, all for 5 min each.
6. Incubate cells or small pieces of tissue in a 50:50 mixture of 100% LR White and 100% ethanol overnight at room temperature and in two daily changes of 100% LR White.
7. Deoxygenate fresh resin under vacuum or by bubbling dry nitrogen gas through it for 10–20 min.
8. Place the cells or tissue in a gelatin capsule or an aluminum weighing boat.
9. Add enough resin to generate a positive meniscus and cover with a piece of Melinex sheet to exclude oxygen.
10. Incubate at 55°C for 24 h to polymerize the resin.
11. Cut thin sections of 50–70 nm and mount on to nickel grids.

CARRY OUT ALL NICKEL GRID INCUBATIONS/RINSES ON DENTAL WAX

12. Incubate sections on drops of PBSG for 10 min.
13. Incubate sections on drops of PBSG containing optimally diluted primary antibodies overnight.
14. Rinse sections on ten 100 µl drops of PBS for 2 min on each drop.
15. Incubate sections on drops of PBSG containing optimally diluted species-specific secondary antibodies conjugated to 10 or 15 nm gold particles at room temperature for 2 h.
16. Rinse sections in ultrapure water for 30–40 s.
17. Counterstain sections by floating grids section side down on drops of 50% (v/v) methanol saturated with uranyl acetate for 0.5–10 min at room temperature, followed by a rinse in 50% (v/v) methanol and a rinse in ultrapure water (44).
18. Counterstain sections by floating grids section side down on drops of lead citrate (45) for 0.5–10 min at room temperature in a Petri dish containing a few grains of moistened potassium hydroxide (to prevent lead carbonate precipitation).
19. Rinse grids extensively in ultrapure water and view at 80 kV in a transmission electron microscope.

Additional note

[f]Used to enhance the retention of phospholipids during primary fixation.

Protocol 3

Immunogold staining following freeze substitution and low temperature embedding, after chemical fixation or after cryoimmobilization

Please refer to the relevant notes in *Protocols 1* and *2*.

Materials and Reagents (please refer to *Appendix 1* for recipes)

- 100 Mesh nickel film grids
- Automated freeze substitution system (Leica Microsystems)
- 1% (w/v) Coldwater fish gelatin[a] in PBS containing 0.001% Tween 20 and Triton X-100[b] (PBSG)
- Dental wax[c] (or parafilm)
- Diamond trim tool and 45° ultradiamond knife (Diatome AG)
- EM UCT ultramicrotome, automated freeze substitution device and CPC (cryo-prep centre) freezing station (Leica Microsystems)
- FEI Tecnai 120 transmission electron microscope
- 4% (w/v) Formaldehyde (made from freshly depolymerized paraformaldehyde) in 0.1 M PIPES buffer (pH 7.4) containing 2 mmol/l calcium chloride[f]
- Lead citrate
- 50:50 Mixture of 100% methanol and 100% HM20 resin[g]
- 50% (v/v) Methanol
- 50% (v/v) Methanol containing saturated uranyl acetate
- 100% Methanol containing 0.05% (w/v) uranyl acetate
- Dulbecco's 'A' PBS (pH 7.6)
- 0.1 M PIPES buffer (pH 7.4)
- Primary antibodies optimally diluted in PBSG
- Secondary antibodies, optimally diluted in PBSG and raised against the species of the primary antibody and conjugated to 10 or 15 nm colloidal gold particles
- 30% (v/v) Polypropylene glycol in PBS (add 1%, w/v, BSA if cells are the subject)
- 0.9% (w/v) Sodium chloride
- Ultrapure water

Method

1. Rinse cells or small pieces of tissue twice in 0.9% sodium chloride.
2. Incubate cells or small pieces of tissue in 4% (w/v) formaldehyde (made from freshly depolymerized paraformaldehyde) in 0.1 M PIPES buffer (pH 7.4) containing 2 mmol/l calcium chloride[f] for 1 h at 4°C. If the cells are adherent, scrape them free from the substrate and transfer to 1.5 ml tubes.
3. Rinse cells or small pieces of tissue four times in 0.1 M PIPES buffer over a period of 20 min and twice in ultrapure water.
4. Incubate cells or small pieces of tissue in 30% (v/v) polypropylene glycol in PBS (add 1%, w/v, BSA if cells are the subject) at room temperature for 2 h.

5. If the subjects are cells, spin down to concentrate them, aspirate off the medium, and transfer them to a small piece of aluminum foil. If the subject is a small piece of tissue, drain it and transfer it to foil.
6. Freeze the cells or small pieces of tissue in liquid propane cooled in liquid nitrogen. If the subject is to be cryoimmobilized without chemical fixation, freeze it by impact against a gold-coated copper block.
7. Transfer cells or small pieces of tissue to the automated freeze substitution system and incubate for 24 h at -90°C in pure methanol containing 0.05% (w/v) uranyl acetate.
8. Warm cells or small pieces of tissue to −70°C and maintain for 24 h.
9. Warm cells or small pieces of tissue to −50°C and rinse in four changes of pure methanol over a period of 2 h.
10. Mix and deoxygenate 100% HM20 resin by bubbling dry nitrogen gas through it for 5 min.
11. Incubate cells or small pieces of tissue in a 50:50 mixture of 100% methanol and 100% HM20 resin at room temperature overnight.
12. Incubate cells or small pieces of tissue in 100% HM20 resin at room temperature for 4 days, changing the resin daily.
13. Polymerize the resin by UV irradiation for 24 h at −50°C, 24 h at −40°C, and 48 h at 15°C.
14. Cut thin sections of 50–70 nm and mount on to nickel grids.

CARRY OUT ALL NICKEL GRID INCUBATIONS/RINSES ON DENTAL WAX

15. Incubate sections on drops of PBSG at room temperature for 10 min.
16. Incubate sections on drops of PBSG containing optimally diluted primary antibodies at room temperature overnight.
17. Rinse sections on ten 100 μl drops of PBS for 2 min on each drop.
18. Incubate sections on drops of PBSG containing optimally diluted species-specific secondary antibodies conjugated to 10 or 15 nm gold particles at room temperature for 2 h.
19. Rinse sections in ultrapure water for 30–40 s.
20. Counterstain sections by floating grids section side down on drops of 50% (v/v) methanol containing saturated uranyl acetate for 0.5–10 min at room temperature, followed by a rinse in 50% (v/v) methanol and a rinse in ultrapure water (44).
21. Counterstain sections by floating grids section side down on drops of lead citrate (45) for 0.5–10 min at room temperature in a Petri dish containing a few grains of moistened potassium hydroxide (to prevent lead carbonate precipitation).
22. Rinse grids extensively in ultrapure water and view at 80 kV in a transmission electron microscope.

Additional note

[9]HM20 is a low-temperature resin, providing low viscosity at low temperatures.

Protocol 4

Immunogold staining of ultrathin thawed cryosections

Please refer to the relevant notes in *Protocols 1* and *2*.

Materials and Reagents (please refer to *Appendix 1* for recipes)

- 400 Mesh nickel film grids
- 1% (w/v) Coldwater-fish gelatin[a] in PBS containing 0.001% Tween 20 and Triton X100[b] (PBSG)
- Diamond trim tool and 45° ultracryodiamond knife (Diatome AG)
- EM UCT ultramicrotome, frozen cryosection module (FCS) and CPC (cryo-prep centre) freezing station (Leica Microsystems)
- FEI Tecnai 120 transmission electron microscope
- 2% (w/v) Formaldehyde (made from freshly depolymerized paraformaldehyde)
- 8% (w/v) Formaldehyde (made from freshly depolymerized paraformaldehyde) in 0.1 M PIPES buffer (pH 7.4) containing 2 mmol/l calcium chloride[f]
- 10% (w/v) Gelatin in PBS
- Hardened filter paper
- 50% (v/v) Methanol
- 50% (v/v) Methanol containing saturated uranyl acetate
- 2% (w/v) Methylcellulose[h] and 3% (w/v) aqueous uranyl acetate in ratios varying from 9:1 to 5:1
- Dulbecco's 'A' PBS (pH 7.6)
- 0.1 M PIPES buffer (pH 7.4)
- Primary antibodies optimally diluted in PBSG
- Secondary antibodies, optimally diluted in PBSG and raised against the species of the primary antibody and conjugated to 10 or 15 nm colloidal gold particles
- 0.9% (w/v) Sodium chloride
- 50:50 Mixture of 2.3 M sucrose and 2% (w/v) methylcellulose[h]
- 1.9 M sucrose[i] and 10% (w/v) polyvinylpyrrolidone (PVP-10)[i]
- Ultrapure water

Method

1. Rinse cells or small pieces of tissue twice in 0.9% (w/v) sodium chloride.
2. Incubate in 8% (w/v) formaldehyde (made from freshly depolymerized paraformaldehyde) in 0.1 M PIPES buffer (pH 7.4) containing 2 mmol/l calcium chloride[f] for 1 h at 4°C. If the cells are adherent, scrape them free from the substrate and transfer to 1.5 ml tubes.
3. Rinse four times in 0.1 M PIPES buffer at room temperature over a period of 20 min and twice in ultrapure water.
4. Incubate in 10% (w/v) gelatin in PBS for 2 h and spin down to form a pellet. Cool to 4°C and fix in 2% (w/v) formaldehyde for 2 h. Small pieces of fixed tissue can be trimmed to 0.5 mm in one dimension.
5. Trim to 0.5 mm^3 cubes and incubate in 1.9 M sucrose[i] and 10% (w/v) PVP-10[i] overnight at 4°C.

6. Freeze the cubes on to sectioning pins in liquid nitrogen. Transfer to the frozen cryosection module and cut thin sections of 90–140 nm.
7. Place a drop of a 50:50 mixture of 2.3 M sucrose and 2% (w/v) methylcellulose in a 1 mm diameter loop and retrieve the sections from the frozen cryosection module. Allow the sucrose to thaw at room temperature and touch the sections on to the Formvar surface of a film grid.
8. Transfer to drops of PBSG.
9. Incubate sections on drops of PBSG containing optimally diluted primary antibodies at room temperature overnight.
10. Rinse sections on ten 100 μl drops of PBS for 2 min on each drop.
11. Incubate sections on drops of PBSG containing optimally diluted species-specific secondary antibodies conjugated to 10 or 15 nm gold particles at room temperature for 2 h.
12. Rinse sections on ten 100 μl drops of PBS for 2 min on each drop.
13. Rinse briefly in ultrapure water and incubate sections on drops of 2% (w/v) methylcellulose and 3% (w/v) aqueous uranyl acetate in ratios varying from 9:1 to 5:1, until the desired contrast is achieved. Blot away excess stain using hardened filter paper and air dry sections before viewing at 80 kV in a transmission electron microscope.

Additional notes

[h]Used as an embedding medium to prevent the sections collapsing totally during air drying.
[i]Used as a cryoprotectant.

Acknowledgements

This chapter is dedicated to my father on the occasion of his 83rd birthday (J.N.S.).

3. REFERENCES

★ 1. **Coons AH, Creech HJ & Jones RN** (1941) *Proc. Soc. Exp. Biol.* **47**, 200–202. – *The first practical demonstration of tagging an antibody with a fluorochrome.*

★ 2. **Singer SJ** (1959) *Nature*, **183**, 1523–1524. – *The first practical demonstration of tagging an antibody with ferratin.*

3. **Rifkind RA, Hsu KC & Morgan C** (1964) *J. Histochem. Cytochem.* **12**, 131–136.

4. **Faulk WP & Taylor GM (1971)** *Immunochemistry*, **8**, 1081–1083.

5. **Frens G** (1973) *Nat. Phys. Sci.* **241**, 20–22.

★★ 6. **Griffiths G** (1993) *Fine Structure Immunocytochemistry*. Springer-Verlag, Heidelberg, Germany. – *A comprehensive description of the ultrathin thawed cryosection technique of immunolabeling.*

★★★ 7. **Lucocq J** (1993) *Trends Cell Biol.* **3**, 354–358. – *A succinct description of quantification of immunolabeling using unbiased stereological techniques.*

★★ 8. **Storm-Mathisen J & Ottersen OP** (1990) *J. Histochem. Cytochem.* **38**, 1733–1743. – *A description of how to correlate gold label density with antigen concentration.*

★★ 9. **Hayat MA** (1981) *Fixation for Electron Microscopy.* Academic Press, NY, USA. – *A thorough description of the chemistry of fixation for electron microscopy.*

★★ 10. **Glauert AM & Lewis PR** (1998) *Practical Methods in Electron Microscopy,* vol. 17: *Biological Specimen Preparation for Transmission Electron Microscopy.* Portland Press, London, UK. – *A thorough description of the chemistry of fixation for electron microscopy and choice of embedding media.*

11. **Bendayan M, Nanci A & Kan FWK** (1987) *J. Histochem. Cytochem.* **35**, 983–996.

★★★ 12. **Peters PJ, Mironov A Jr, Peretz D, *et al.*** (2003) *J. Cell Biol.* **162**, 703–717. – *A superb example of the use of the ultrathin thawed cryosection technique of immunolabeling.*

13. **Skepper JN, Woodward JM & Navaratnam V** (1988) *J. Mol. Cell. Cardiol.* **20**, 343–353.

★★★ 14. **Crapo JD, Oury T, Rabouille C, Slot JW & Chang L-Y** (1992) *Proc. Natl. Acad. Sci. U. S. A.* **89**, 10405–10409. – *An example of the apparently paradoxical requirement to use strong fixation to retain and label soluble antigens.*

15. **Tokuyasu KT** (1986) *J. Microsc.* **143**, 139–149.

★★★ 16. **Liou W, Geuze HJ & Slot JW** (1996) *Histochem. Cell Biol.* **106**, 41–58. – *A review of modifications to the ultrathin thawed cryosection techniques that significantly improve ultrastructural preservation.*

17. **Chung D, Kim E & So PT** (2006) *Opt. Lett.* **31**, 945–947.

18. **Egner A, Verrier S, Goroshkov A, Söling H-D & Hell SW** (2004) *J. Struct. Biol.* **147**, 70–76.

19. **Willig KI, Rizzoli SO, Westphal V, Jahn R & Hell SW** (2006) *Nature* **440**, 935–939.

★ 20. **Brændgaard H & Gundersen HJG (1986)** *J. Neurosci. Meth.* **18**, 39–78. – *An example of the 'reference trap' that can invalidate quantification.*

★★★ 21. **Howard CV & Reed MG** (1998) *Unbiased Stereology.* BIOS Scientific Publishers, Oxford, UK. – *A detailed description, with practical examples, of how to apply stereology.*

★ 22. **Gundersen HJG** (1977) *J. Microsc.* **111**, 219–223. – *The forbidden line rule.*

23. **Lucocq JM, Habermann A, Watt S, Backer JM, Mayhew TM & Griffiths G** (2004) *J. Histochem. Cytochem.* **52**, 991–1000.

★ 24. **Mayhew TM, Lucocq JM & Griffiths G** (2002) *J Microsc.* **205**, 153–164. – *Nonparametric methods of quantifying immunogold labeling.*

★★★ 25. **Skepper JN** (2000) *J. Microsc.* **199**, 1–36. – *A comprehensive review of the methods available for immunolabeling and their strengths and weaknesses.*

26. **Causton B** (1984) In: *Immunolabelling for Electron Microscopy,* pp. 29–36. Edited by JM Polak & IM Varndell. Elsevier, Amsterdam.

27. **Bendayan M & Zollinger M** (1983) *J. Histochem. Cytochem.* **31**, 101–109.

28. **Probert L, de May J & Polak J** (1981) *Nature,* **294**, 470–471.

29. **Bendayan M, Nanci A, Herbener GH, Gregoire S & Duhr MA** (1986) *Am. J. Anat.* **175**, 379–400.

30. **Varndell IM, Sikri KL, Hennessy RJ, *et al.*** (1986) *Cell Tissue Res.* **246**, 197–204.

31. **Newman TM, Severs NJ & Skepper JN** (1991) *Cardioscience,* **2**, 263–272.

★ 32. **Causton B** (1981) *Proc. R. Microsc. Soc.* **16**, 265–271. – *The introduction of LR White.*

★★ 33. **Berryman MA & Rodewald RD** (1990) *J. Histochem. Cytochem.* **38**, 159–170. – *A method for improving membrane preservation during embedding in acrylic resin.*

34. **Zajicek J, Wing M, Skepper J & Compston A** (1995) *Lab. Invest.* **73**, 128–138.

★★ 35. **Monaghan P & Robertson D** (1990) *J. Microsc.* **158**, 355–363. – *A method for cryoimmobilization, freeze substitution, and low-temperature embedding without chemical fixation.*

36. **Skepper JN, Whiteley A, Browne H & Minson A** (2001) *J. Virol.* **75**, 5697–5702.

★ 37. **Tokuyasu KT** (1973) *J. Cell. Biol.* **57**, 551–565. – *An early review on the ultrathin thawed cryosection techniques by its pioneer.*
38. **Tokuyasu KT** (1983) *J. Histochem. Cytochem.* **31** (Suppl. 1A), 161–167.
39. **Tokuyasu KT** (1986) *J. Microsc.* **143**, 139–149.
40. **Tokuyasu KT** (1989) *Histochem. J.* **21**, 163–171.
41. **Liou W, Geuze HJ, Geelen MJH & Slot JW** (1997) *J. Cell. Biol.* **136**, 61–70.
42. **Klumperman J, Kuliawat R, Griffith JM, Geuze HJ & Arvan P** (1998) *J. Cell. Biol.* **141**, 359–371.
43. **Russell FD, Skepper JN & Davenport AP** (1998) *Circ. Res.* **83**, 314–321.
44. **Gibbons IR & Grimstone AV** (1960) *J. Biophys. Biochem. Cytol.* **7**, 697–716.
45. **Reynolds ES** (1963) *J. Cell Biol.* **17**, 208–212.

CHAPTER 8

Image capture, analysis, and quantification

Jiahua Wu, Anthony Warford, and David Tannahill

1. INTRODUCTION

It is clear that high-throughput immunochemical staining technologies are having a significant impact on biomedical science and with this has come the desire of the scientific community to make large image datasets available for viewing over the internet. Some understanding of how image capture and analysis, as well as rudimentary databases, operate can only serve to make the use of these databases more effective by the end user. We hope that this chapter introduces some basic concepts that underlie the developments in this new and exciting era.

Over the last two decades, the advent of reliable and sensitive methods for determining the localization of proteins and other markers within tissue sections has provided a rich seam of fundamental expression data that have underpinned the enormous success in modern clinical diagnostics and biomedical research. However, the recent application of robotics to the immunohistochemical staining laboratory now means that even small laboratories can generate vast quantities of data in a relatively short period of time. With such a wealth of immunohistochemical staining data, there are many new challenges to be faced, particularly in accurately imaging immunohistochemical staining data. By and large, the immunohisto-chemical staining laboratory has moved away from traditional film-based imaging technologies as the price and capability of digital solutions has become more affordable. As well as benefits in speed and capacity, digital imaging opens up a whole new world of analytical possibilities to process immunohistochemical staining expression data. However, the capture and analysis of large numbers of digital images brings with it a number of logistical considerations for the storage and handling of images across computer networks. Significant challenges also exist in the retrieval of

Immunohistochemistry: *Methods Express* (S. Renshaw, ed.)

images and their subsequent display, when individual requirements can range from small numbers of research scientists accessing a single workstation to large groups of clinicians spread over several geographically distant sites requiring simultaneous access. In this regard, the ultimate challenge may be represented by the large open-access reference expression databases that are currently under development, which expect thousands of web-hits per day. Examples of such expression databases include: www.brainatlas.org; www.genepaint.org; www.hpr.se; http://genex.hgu.mrc.ac.uk; www.ncbi.nlm.nih.gov/projects/gensat/; www.stjudebgem.org; and www.eurexpress.org.

Whilst the methods of immunohistochemical staining are firmly established and appear to evolve at a steady rather than revolutionary rate (see Chapter 1), recent progress in the automation of immunohistochemical staining processes (see Chapter 10), coupled with the application of new imaging sensors and powerful computing hardware, have made projects feasible that were simply impractical a few short years ago. It is now apparent that we are entering a decade where the interface of engineering, computing, and biology holds enormous promise for understanding the processes of biology and disease. It is our belief that the creation and analysis of large sets of expression data are pivotal to this.

This chapter describes the key considerations when establishing a modern imaging system for immunohistochemical staining. Because of the rapid pace of technical advances, the plethora of imaging and software platforms, and the vast number of different laboratory scenarios, it is not possible to provide a set of definitive protocols along the lines of that given in a typical chapter of a *Methods Express* book. Instead, we provide a detailed appraisal of the various methods and approaches in current use. In doing so, we do not endorse any particular manufacturer or product; thus, any such reference is purely for illustrative purposes.

In choosing any imaging and microscopy platform, one should be fluent in the appropriate terminology and readers are therefore encouraged to consult educational primers such as that produced by Olympus (see http://www.olympusmicro.com/primer/index.html).

2. METHODS AND APPROACHES

2.1 Image capture

2.1.1 Hardware and software

With regard to image capture hardware, the key rule is to obtain the highest-quality images in the first place. For both image analysis and image processing, attention has to be paid to errors occurring during the image formation and acquisition. For image analysis, which has an image as input and data as output, the quality may not always be critical. Some measurements might be done on noisy images or on nonsharp images without losing accuracy. However, for image processing, image quality has to be as high as possible since the output is ultimately an image. Effort invested in obtaining high-quality images is rewarded with less trouble in subsequent procedures.

In order to choose hardware for image capture, careful thought is needed to determine the laboratory's imaging requirements. The first question must be: is a fluorescent imaging system required? The addition of fluorescent capabilities adds significantly to system cost and design, both in the microscope and in sensors/software. Obviously, confocal microscope systems offering superb image detail are available (see Chapter 6). Here, we have mainly limited ourselves to imaging of the most prevalent type of immunohistochemical staining experiment in which primary antibody binding to an antigen in tissue sections is visualized by the reaction of chromogenic substrates (e.g. from primary, secondary, or tertiary reagents coupled with alkaline phosphatase or horseradish peroxidase – see Chapter 4) on compound (wide-field) light microscopes. The use of high-contrast chromogenic substrates provides high sensitivity with good signal-to-noise ratios. However, inherent in some of these methodologies are the nonlinear enzyme reaction rates, as well as the restricted color space occupied by the chromogenic reaction products, which can make absolute signal quantification more problematic than similar fluorescence experiments. Such experiments may be viewed as semi-quantitative at best, but, with good experimental design and the application of appropriate analysis tools, these problems can be minimized.

In assessing the microscopy requirements, it must be determined whether microscope motorization is beneficial. All major microscope suppliers offer motorized platforms in which a computer can control every aspect of the microscope from illumination to stage control. Motorization is not simply a convenience, but, in conjunction with appropriate microscope control software, it is essential for some applications. For example, if one is trying to produce a composite image of a large tissue section, then the precision that a motorized stage affords is essential, as it is for any technique that requires fine control in the *z*-axis. Microscope automation is also desirable if the same sample is to be imaged at different magnifications, with different filter sets or exposures. Two further considerations should be borne in mind. First, one must be cautious when transferring image data between devices, especially between different manufacturers, because an image can be represented as a 'visual' image in many ways. Also, various devices can introduce different sources of nonlinearity into the image data (1). Secondly, a network with high bandwidth (100–1000 megabits) is crucial for transferring large images, especially in the high-throughput environment.

Most microscope manufacturers supply image capture software that offers a bewildering array of tools for image capture and for analysis and data management. Often these packages are modular and only the tools appropriate to a particular use need be purchased. However, it should be remembered that stand-alone software can often provide support for a more extended range of cameras and other hardware. Such packages often include sophisticated programming and image analysis tools (e.g. METAMORPH from Molecular Devices Inc., IPLAB from Scanalytics and OPENLAB from Improvision). The following features should be a part of any imaging system:

- A fast and simple-to-use interface.
- The ability to acquire multiple color channels.
- The ability to store raw images as well as processed images.
- The capability to support different cameras, filters, and stages.
- Comprehensive image-enhancement tools.
- On-screen previews.
- The capability to batch capture.
- Scripting for automating and customizing.
- The ability to export images to other programs.

2.1.2 Methods for digitizing images

Image acquisition

A conventional compound microscope uses a well-established design that comprises a series of lenses and aperture diaphragms to allow the study of fine tissue details based on different light transmission properties. The deposition of chromogenic reaction products in immunohistochemical staining experiments therefore allows the visualization of specific structures. To capture immunohistochemical staining images on a compound microscope, some sort of sensing device is used, which may be film or digital. The vast majority of imaging systems in use today are still based on this fundamental set-up. More recently, however, novel devices have become available that are not based on traditional microscope and imaging sensors. These systems, such as the Aperio SCANSCOPE range or the NANOZOOMER by Hamamatsu, may offer additional capabilities, particularly in the scanning and analysis of large tissue sections. Recently, DMetrix, Inc. have produced an extremely rapid slide scanning system that employs many innovations including a novel compound lens akin to an insect eye. This DX-40 system is capable of scanning a whole slide at 20× magnification in under two minutes.

In the high-throughput environment, automation of all aspects of the imaging system is required (also see Chapter 10). Systems are now available where multiple slides are accommodated in a slide 'hotel' and then loaded in sequence on to an automated imaging platform e.g. ARIOL SL-50 (Applied Imaging), the .SLIDE SYSTEM (Soft Imaging System), SCANSCOPE T2 (Aperio). Such hardware places special demands on the capabilities of the imaging software and hardware, many of which can be illustrated in the ARIOL SL-50 system with its 50-space slide hotel. First, each slide is given a bar code and then scanned to produce a record unique to the slide. Bar coding is essential for data tracking into the final database. The record created by bar coding also contains the key metadata such as sample details, clinical data, etc. and represents the link that keeps the image correctly associated with its data in the final database. Bar coding can also be used to instruct the imaging system what to do with the slide. An example protocol for a large tissue section would be to first scan the entire slide at low magnification in order to 'find' the tissue. This would then be followed by a step that instructs the microscope to re-scan only the tissue and not blank slide areas at high power. Even such a simple protocol requires sophisticated software to control exposure, to produce accurate focus across the whole section, and to join different image frames into a composite. Protocols can also be set up to perform any post-processing steps such as signal detection, rare event counting, or quantification. Images and data are then deposited in a dedicated database. Essentially, this is a virtual microscope environment where images and data are called up from a database and viewed and interacted with on screen in a manner similar to using a microscope. At the point of review, further information, such as pathology scoring, annotation, etc., can be added to the record. Whilst the manufacturers of such products provide end-user database solutions that allow the viewing of the images in a virtual microscope environment, it should be remembered that extracting or linking these images and data from these proprietary systems to other database types (e.g. genome browsers) may not be straightforward. It cannot be overemphasized that capturing a large number of large images requires the availability of large volumes of storage (many terabytes). It is easy to overlook this. Image compression using JPEG2000 or other tools may help (see below), but for large numbers of images this will require dedicated computing resources. However, unless the original image is discarded (which may harm subsequent analysis steps), this only adds to the storage requirements. Another key point is that, to retrieve the final image from the database, a high-bandwidth network is required; otherwise it will take time to load images on screen.

Whilst the prediction of the subcellular localization sites of proteins from their amino acid sequences is a fairly long-standing problem in bioinformatics (2), the actual demonstration of subcellular sites of protein localization generally requires the use immunohistochemical staining experiments, particularly with fluorescent secondary reagents (see Chapters 4 and 5). In these scenarios, a compound microscope equipped with fluorescence capabilities is required. By using appropriate excitation and emission filters, an image is formed by the emitted fluorescence and not by the original illumination. Detection of specimen detail does not depend on the size of the detail, but on the amount of fluorochrome attached to it. This method is very sensitive and results in high-contrast images. Fluorescence is not only suitable for the qualitative assessment of signal localization; it also affords benefits in signal quantification. Increased confidence in assigning localization can be achieved in co-localization experiments where output fluorescence is assessed with respect to the signal from a known reference marker (see Chapter 5). A disadvantage of both the conventional and fluorescence microscope is that light originating from the out-of-focus planes is also collected in the imaging device, thus degrading the image and limiting the resolution. The design of a confocal microscope overcomes these limitations to produce extremely high-resolution images (see Chapter 6).

Charge-coupled devices

The most common imaging device used in immunohistochemical staining is the charge-coupled device (CCD) camera. To capture a color image, an image must be acquired for each color channel (red, green, and blue; RGB). The final image is then formed by merging the channels together. Instrument noise can be reduced by taking background images, which are then subtracted from the desired image. There are a variety of ways to generate each color channel – there can be arrangements of colored masks on the CCD sensor, or colored filters can be rotated into the primary optical path for each channel. In some cases, it can be useful to use tunable crystal filters, as these do not require the physical movement of different pieces of glass into the optical path. If a CCD is also to be used for fluorescence, then it is imperative that the CCD sensor has no RGB color masks; otherwise there will be a severe loss in sensitivity. Also, many CCD sensors have an infrared filter that may make detection of longer wavelengths (e.g. Cy3 and Cy5) problematic. Such 'scientific-grade' CCD sensors are built to very high specifications, having very few dead pixel elements. It is also desirable to cool the sensor, as this helps to eliminate background noise from the sensor itself. In some cases, the spectral sensitivity of the sensor may also be important if accurate quantification is required. For example, many sensors are not as efficient at detecting near-infrared compared with visible wavelengths. For ultimate sensitivity (e.g. photon counting), one must consider systems that employ photomultiplier tubes. As is the case with film cameras, the size of the sensor determines how much area can be imaged in one frame. Generally, for large areas several frames need to be taken and reassembled to generate a composite image (see below).

It should be noted that image acquisition (digitization) always produces quantization noise. As modern cameras capture at 12 or 16 bits (65 536 intensity levels) per channel, quantization noise is generally negligible. However, if the signal occupies only a small portion of this dynamic range, severe quantization noise might occur. Even though images are often captured at 12 or 16 bits per channel, for the purposes of visual display, these images are generally displayed as 8 bits per channel and therefore must be compressed to generate display 24-bit composite RGB images (3). This has great potential to introduce artifacts and loss of image detail.

2.1.3 Tissue microarrays

Tissue microarrays (TMAs) are constructed by removing small cylindrical samples (core biopsies) from tissue specimens and arraying these, at high density, into a recipient paraffin block (see Chapter 4, section 9, for a discussion of their use and construction). A major advantage of TMA technology is that a single immunohistochemical staining experiment carried out on a single TMA slide can yield information on the molecular characteristics of hundreds of donor specimens (4). TMAs hold great promise for revolutionizing the analysis of protein distributions in immunochemical staining experiments, similar to the impact that DNA microarrays had on functional genomics (5). However, analyzing TMAs is less straightforward than that of DNA microarrays.

The greatest challenge in TMA techniques is collecting and interpreting the vast amount of data. Whilst commercial systems can automate collection and archiving of the data, they generally only provide limited tools for data/image analysis. For example, the ARIOL SL-50 system provides a robust solution for TMA image capture and uses a software tool to extract the TMA core position from an image. However, such instruments do not replace the interpretive skills of a pathologist/scientist. Manual interpretation of TMA cores constitutes a significant bottleneck in high-throughput operations. For example, it is estimated that the analysis of 30 000 genes in 1000 tumors would take 20 pathologists a year to complete (6). Furthermore, manual interpretation does not allow the accurate quantitative analysis of staining to be undertaken (5). It is also physically impossible for an individual to monitor the protein distributions accurately on several hundreds of thousands of individual tissue cores, never mind conducting statistical analysis and complex group comparisons. Whilst automated software for immunohistochemical staining interpretation is becoming available, it is only in specific cases that it has been shown to be superior to manual methods (e.g. rare-event detection for micrometastatic cells). The challenge for the future is therefore the automated collection and reliable interpretation of TMAs with quantitative output (5).

2.1.4 Three-dimensional imaging

It is now becoming more common to investigate protein distribution in three dimensions (3D) within an organ or tissue. A number of technologies are emerging for the direct 3D imaging of whole stained samples, such as optical projection tomography (7). Understanding gene expression in 3D raises additional questions in both image display and analysis. For further details on optical projection tomography, see http://genex.hgu.mrc.ac.uk/OPT_Microscopy/optwebsite/frontpage/index.htm. A commercial OPT scanner has recently become available from www.bioptonics.com.

It should be noted that a variety of tools exists for the extraction of 3D spatial information from tissue sections. Generally, this is done by taking a number of images at different focal planes in the *z*-axis. These image stacks are then rendered mathematically to produce voxels from which 3D images and iso-surfaces can be created and freely rotated. Most major microscope manufacturers offer such software capabilities, as do stand-alone products such as IMARIS from BitPlane AG and ANALYSIS from Soft Imaging Systems. These applications work best in fluorescence and give superb results with confocal microscopes. However, diffraction grating-based attachments for compound microscopes, such as the APOTOME produced by Zeiss, or mathematical techniques such as deconvolution (e.g. AUTODEBLUR from AutoQuant) (see Chapter 6, section 2.10.3) may begin to offer similar capabilities that compensate for out-of-focus information on compound microscopes. Deconvolution and other image-processing algorithms, such as UNSHARPEN MASK found in ADOBE PHOTOSHOP, also offer the potential to improve image quality in chromogenic immunohistochemical staining experiments, but caution should be exercised in their use so that artifacts are not introduced into the image. A complementary technique that is likely to find increasing use in immunohistochemical staining experiments is extended focal imaging, which attempts to overcome the limited depth of field in images taken on standard wide-field microscopes (e.g. part of the ANALYSIS package). The optical design of the objectives for a compound microscope means that high-resolution 2D images only have a very shallow depth of field. The extended focal imaging software tool overcomes this limitation by collecting only the focused parts from a series of images taken along the *z*-axis (under the control of a motorized stage) to produce a single image with unlimited depth of field. This technique may increase the amount of data available for analysis without the need for additional experiments.

2.2 Image analysis and quantification

Most of what follows is generally applicable to immunohistochemical staining and *in situ* hybridization experiments. *Fig. 1* illustrates the typical steps in image analysis required to estimate the expression levels of a protein in an immunohistochemical staining (TMA) experiment. In attempting to analyze images, many researchers are still inclined to inspect images manually instead of relying on automated image analysis. Unfortunately, different researchers also have different standards in judging features of interest, and these 'intuitive' standards are also subject to individual user variability. As a result, the manual assessment of large numbers of images is not sufficiently objective to maintain quality and consistency among experiments and across different laboratories. There is an urgent need, therefore, for a rigorous definition for signal characterization; the development of robust automatic image analysis algorithms could fill this gap. An automated approach is also necessary for the potential benefits of quantitative classification to be fully realized. A completely automated approach would not be limited in the number of images that can be evaluated and would offer advantages in speed and reproducibility that are difficult to achieve manually. Even the simple, full automation of the image-processing stage would be of great value, and, whilst many attempts have been made, most have drawbacks making their full-scale utilization prohibitive. There is no standard rule in this field.

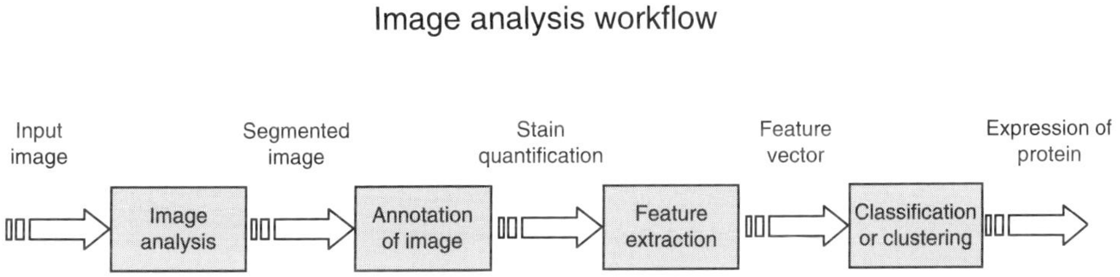

Figure 1. A typical image analysis pipeline in an immunohistochemical staining experiment.

2.2.1 Correcting imaging defects

Many factors can impact the image quantification procedure. For example, effects due to image background, signal-to-noise ratio, feature imaging response and saturation, experimental design, and execution all must be taken into account. The first steps in image processing are designed to eliminate noise in the image data. This includes both small random specks and trends in background values. Dust and other defects on the slide surface produce spike noise, i.e. small spots of high intensity. In addition, microscopic scratches or small pieces of dust produce bright snake-like distortions (8). The sources of noise can also be divided into instrument noise, which is produced by the imaging system itself, and experimental noise, which results either from aberrations in the slide or from nonspecific antibody binding to the slide. Global background correction uses the same constant to represent the background across a slide. This constant can be calculated by using a negative-control slide. However, global estimation does not take into account the local background variations that exist in many cases. Local background correction provides greater flexibility than global correction (9, 10). In order to illustrate the principles of noise reduction, we describe the operation of three filters below, but it should be remembered that other filters, such as wavelet thresholding (11) and morphological filters (12, 13), may also be used. Note that if an image is also to be processed with a thresholding step, one can filter for noise before and after thresholding to determine which is most effective.

Median filter

One of the simplest ways to remove noise is to smooth the image using a nonlinear median filter. This simply adds together the pixel brightness values in each small region of the image, divides by the number of pixels in the neighborhood, and then uses the resulting values to construct a new image (13). *Fig. 2(A)* (also available in the color section) illustrates a TMA image before and after the application of a 3×3 median filter, from which we see that most of the noise has been removed.

Top-hat filter

Yang *et al.* (2002) (10) proposed the top-hat filter to remove uneven background across a sample (see Chapter 6, section 2.10.2). This filter is an example of a suppression operator. It removes pixels from the image if they do not meet some criteria of being 'interesting' and leaves alone those pixels that do. It can be used to remove small defects by replacing the pixels with values taken from the surrounding neighborhood (see *Fig. 2B*, for example, also available in the color section). For further details, see (13).

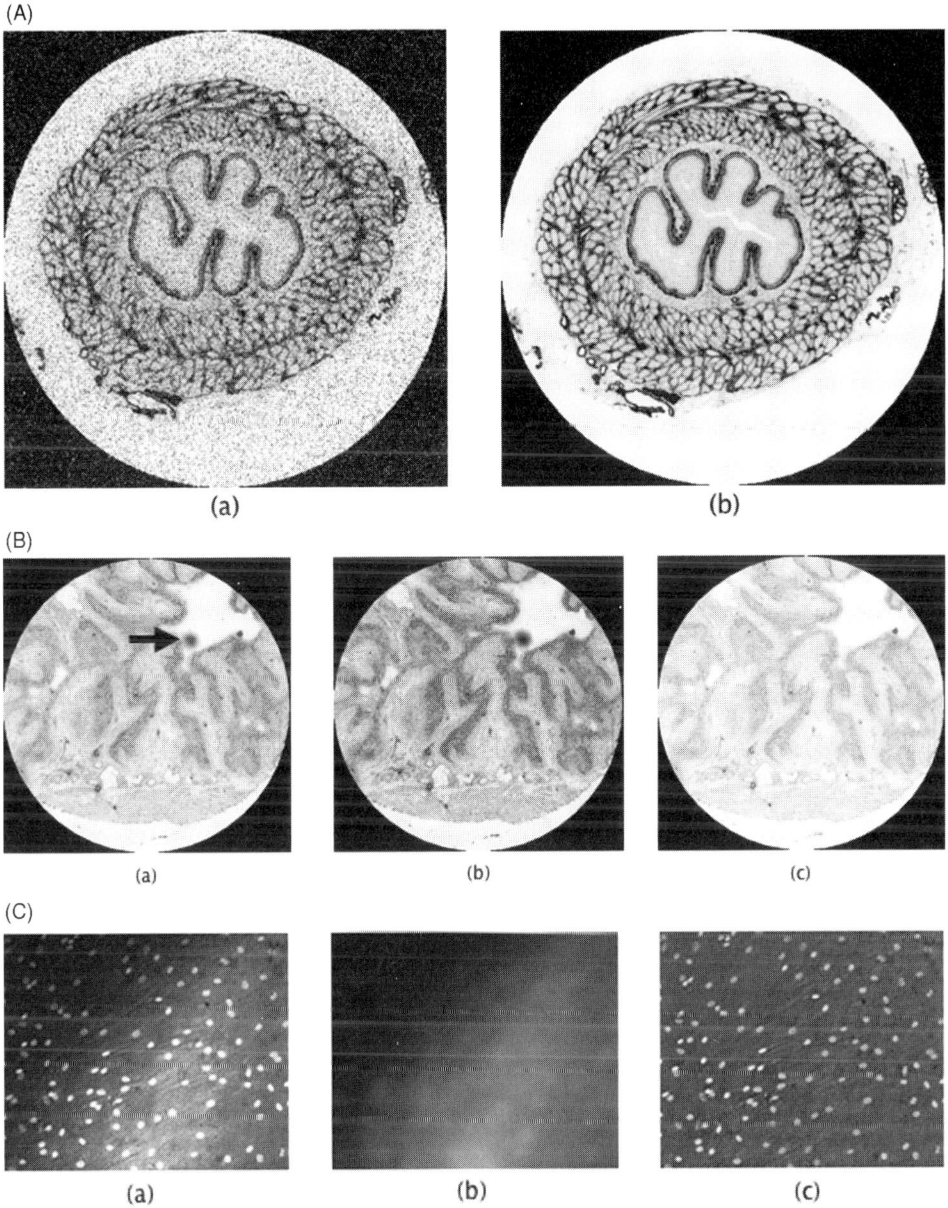

Figure 2. Correction of imaging defects (see page xix for color version).
(*A*) Using a median filter. (*a*) A TMA core image displaying (artificial) background noise. (*b*) Median-filtered image of (a) showing noise removal. For the purposes of printed display, the noise in image (*a*) has been exaggerated. (*B*) Using a top-hat filter. (*a*) A TMA core image with background noise, which includes a large spot artifact (arrow). (*b*) Image showing color intensities of the image in (*a*). (*c*) Top-hat-filtered image showing the removal of the spot artifact. (*C*) Using a low-pass filter. (*a*) A field of nuclei with uneven illumination. (*b*) Estimated background image generated from a low-pass filter on (*a*). (*c*) Correction of uneven illumination by subtracting (*b*) from (*a*).

Low-pass filter

A low-pass filter can be used to remove large-scale image background variations such as illumination variations across the field of view (3, 14). Illumination correction exploits the property that illumination variations change at a much lower rate than features do – illumination variation involves only low spatial frequencies. An estimate of the background may be found by removing all but the low spatial frequencies from the image. This can be achieved using a low-pass filter. The background estimate can then be subtracted from the original image to correct for background variation. *Fig. 2(C)* (also available in the color section) shows a low-pass filter correction of an image of nuclei with uneven illumination.

2.2.2 Color space

The color space in which an image is collected also has important implications for the image quantification procedure. CCD color cameras capture images based on variable combinations of three primary colors, red (R), green (G), and blue (B), by measuring the intensity of transmitted light for each pixel in three wavelength ranges by the use of R, G, and B filters. The resulting R, G, and B intensities together may be regarded as a point in a 3D Euclidean coordinate space (3) (see *Fig. 3A*, also available in the color section). The individual R, G, and B intensities reflect both the absorption characteristics of the stain and the amount of stain present. Variation in the amount of stain bound to different locations of the specimen will result in variations in the intensities for different pixels, even if only a single stain is present. In other words, differential staining in a sample will result in a nonlinear relationship between the measured intensities at different locations in the sample. Within the RGB space, these intensity variations will comprise a complex shape 3D domain for each stain. Stain recognition is possible by classifying each

Figure 3. RGB and HSI color space (see page xx for color version).
(*A*) RGB color space. (*a*) RGB color space can be geometrically represented in a 3D domain, where the coordinates of each point represent the values of the red, green and blue components, respectively. (*b*) RGB color cube. (*c*) Color image of a TMA core made up of a red component (*d*), a green component (*e*), and a blue component (*f*). (*B*) HSI color space. (*a*) Bi-conic representation of HSI color space. (*b*) Color image of a TMA core. (*c*) Hue component extract from (*b*). This is considered to be an angle between a reference line and a color point in HSI cone space. The range of hue values is from 0 to 360°. (*d*) Saturation component extracted from (*b*). This represents the radial distance from the cylinder centre; the nearer the point is to the center, the lighter the color. (*e*) Intensity component extracted from (*b*). This is the height in the *z*-axis direction. The axis of the cylinder describes the gray levels where zero intensity (minimum) is black and full (maximum) intensity is white.

(A)

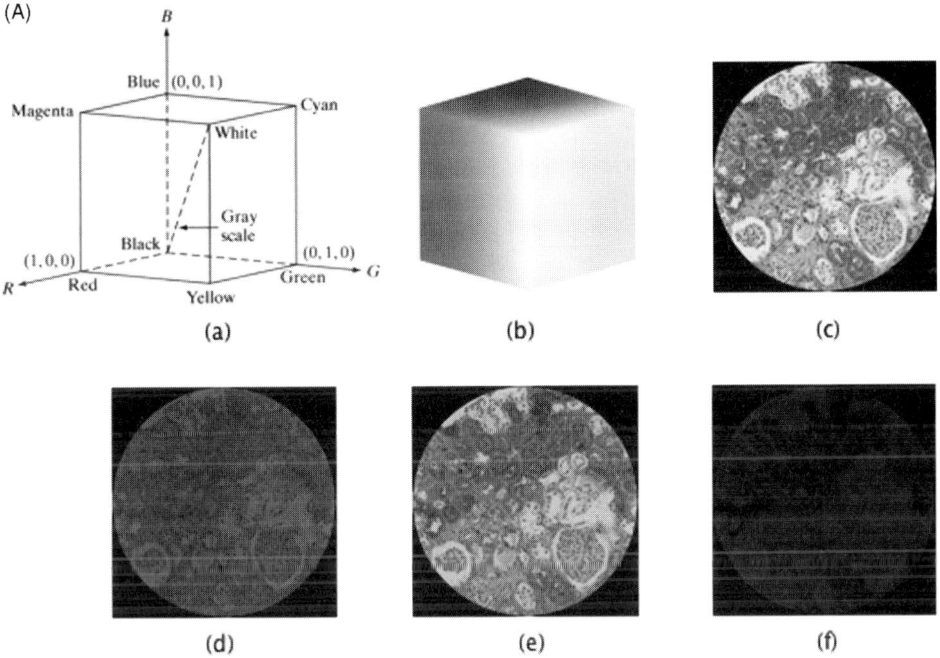
B
Blue (0, 0, 1)
Magenta
Cyan
White
Gray
scale
Black
(1, 0, 0)
(0, 1, 0)
R
Red
Yellow
Green
G
(a)
(b)
(c)
(d)
(e)
(f)

(B)

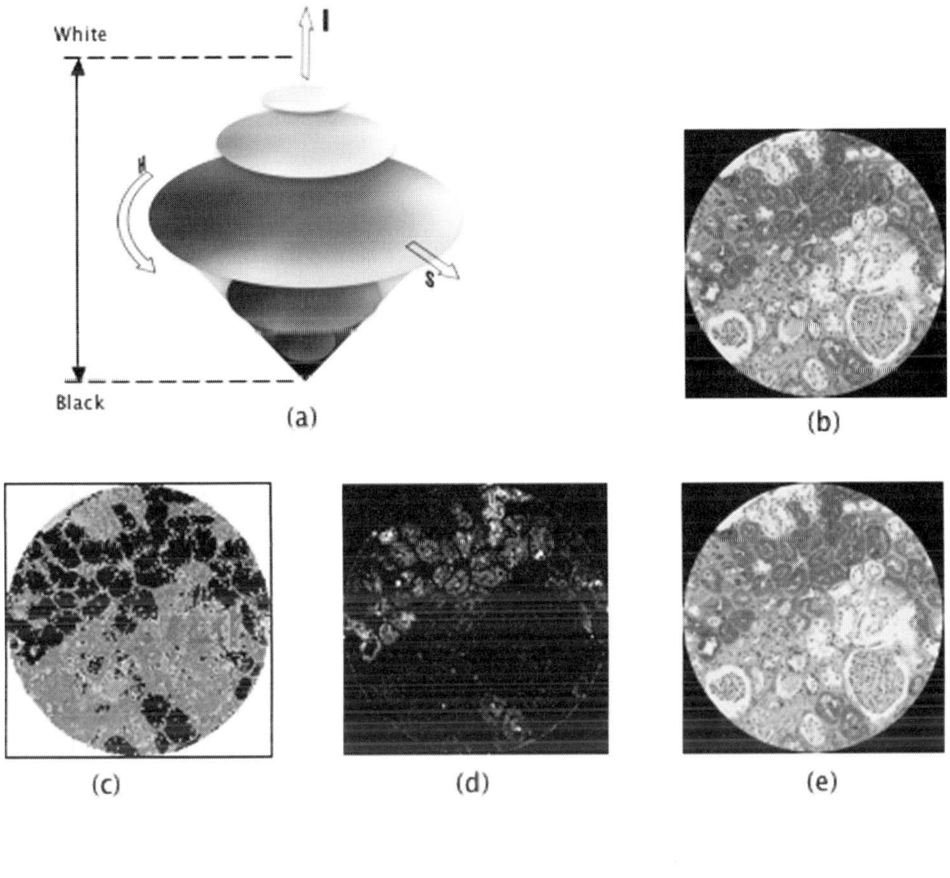
White
I
H
S
Black
(a)
(b)
(c)
(d)
(e)

pixel to the stain corresponding to the 3D domain in which the RGB point is located.

RGB color space is suitable for color display, but is less good for image analysis and segmentation because of the high correlation among the R, G, and B components. This is problematic because the information of interest, i.e. the color of the stain (determined by the absorption characteristics), is mixed with variations in the amount of stain. This means that any operations that process intensities (e.g. illumination correction) can adversely affect image color. Furthermore, in comparing two stains, if the absorption characteristics are not sufficiently different within the three sensitivity bands of the camera, then processing based on the RGB data will be difficult (15).

To overcome the limitations that RGB color space imposes on image analysis, an alternative method of representing color using HSI space can be employed. In this model, light is separated into intensity and chromaticity components. Intensity (I) and is dependent on the amount of energy received but independent of the color of the light. Chromaticity is expressed as hue (H) (i.e. the dominant wavelength of the light) and saturation (S) (i.e. color purity; for example, pink is red with a low degree of saturation). The HSI color space is geometrically more complex than RGB space (compare *Fig. 3A* and *B*, also available in the color section). In HSI color space, classification of pixels is based on two dimensions, hue (dominant wavelength) and saturation (purity of the color). These are independent of the signal intensity (which is represented in the third dimension). To convert from RGB to HSI space, the mathematical formulae are considerably complex and the reader is referred to (14) for further details.

The use of HSI space allows algorithms based on gray levels to process the intensity component of the HSI description. Furthermore, to separate (i.e. segment) objects with different colors, algorithms can be applied to the hue component only. This is especially useful when the images have uneven illumination, as hue is independent of intensity values. Hue is particularly useful in cases where the illumination level varies from point to point or image to image. It should be noted, however, that it is difficult to transform these results back into RGB values, as RGB encoding requires hue, saturation, and intensity values to be present. *Table 1* summarizes the main features of HSI and RGB color space.

2.2.3 Image segmentation

After noise removal, the next step in image processing is segmentation. This is the extraction and separation of different features of an image and is a fundamental task in the analysis of immunohistochemical staining images. This can be used to define features for further quantitative

Table 1. Characteristics of HSI and RGB color space

Color space	Advantages	Disadvantages
RGB	Convenient for display	Not good for color image processing
HSI	Similar to human color perception Useful for cases where the illumination level varies, as hue is invariant to certain types of highlights, shading, and shadows Hue can be useful for separating objects with different colors	Difficult to display Conversion to RGB for visual display can cause artifacts due to nonlinear transformation

analysis and also to correct image defects. For a review of the methods and application of color segmentation, see (16). A variety of approaches have been applied to immunohistochemical stained images and these can be divided into three main groups (17): pixel-based techniques, such as clustering; area-based techniques, such as split-and-merge algorithms; and edge detection, including the use of color-invariant snakes. Some examples of segmentation principles are given below.

Adaptive histogram thresholding

Thresholding is a very simple and often-used method for image segmentation and is based on histogram characteristics of the pixel intensities of the image. For an overview of thresholding techniques, see (18) and (19). In order to obtain a satisfactory segmentation result by thresholding, a uniform background is required. Many background correction techniques exist, but they may not always result in an image that is suitable for thresholding. The transition between object and background may be diffuse, making an optimal threshold level difficult to find. In addition, a small change in the threshold level may have a great impact on later analyses.

Adaptive local thresholding can be used to circumvent the problem of varying background, or as a refinement to coarse global thresholding (15). In essence, images are divided into small overlapping regions and each region is tested for bimodality by trying to fit a pair of Gaussian curves to the histograms (20). If a region does not exhibit bimodality, it is assumed to be all object or all background and is assigned a threshold from a neighboring region. Such thresholding algorithms are nonparametric, meaning that they operate independently of the actual distribution in intensity values. The main drawback in such histogram-based methods is that they do not take shape information into account and the outcome can be unpredictable, especially in cases of low signal-to-noise ratios. In addition, spike noise or contamination from other spots may be classified into the spot, leading to errors in the estimated intensity values.

Another thresholding problem that can be addressed by adaptive histogram thresholding is the accurate identification of individual TMA cores on a slide. This involves finding a circle that separates out the core from the background and uniquely separates the cores. This task consists of three steps. The first is background equalization for intensity variation to remove any smooth variations across the grid drawn over the TMA. Next, statistical intensity modeling and optimum thresholding of the grid is performed. When a core is present, the intensity distribution of the pixels within the grid is modeled using a Gaussian-mixture model (20). Finally, after thresholding and segmentation, a best-fit circle is computed for the final core segmentation.

Edge detection

Determining the edges of a feature in an image (e.g. a cell or nuclear membrane) is a basic requirement of immunohistochemical staining image analysis. Edge detection is extensively utilized for gray-level image segmentation and is based on the detection of discontinuity in gray levels, trying to locate points with abrupt changes. There are many types of parallel differential edge filters such as Canny, Laplacian, Roberts, Sobel, Kirsch, and Prewitt filters (3, 14, 15). The Sobel filter is one of the most useful and widely available edge filters (15). See *Fig. 4(A)* (also available in the color section) for examples of a TMA core processed with different edge detection filters.

Seeded region growing

The aim of the segmentation methods described above is to find borders between regions; the following methods construct regions directly. It is straightforward to construct regions from their borders and, conversely, it is straightforward to detect borders of existing regions. However, edge-based methods and region-growing methods do not usually give the same

Figure 4. Image segmentation using edge detection and watershed segmentation (see page xxi for color version).
(*A*) Use of edge detection filters. Original TMA core image (*a*) processed with Canny (*b*), Sobel (*c*), Prewitt (*d*), Lapacian (*e*) and Roberts (*f*) edge detection filters. (*B*) The principle of watershed segmentation. (*a*) A binary image of overlapping circles. (*b*) Euclidean distance transformation of the binary image in (*a*) with pixels color coded to show distance from boundary. A watershed is the ridge that divides areas drained by different river systems. A catchment basin is the geopgraphical area draining into a river or reservoir. (*c*) Result of watershed segmentation of (*a*). (*d*) Original image of a field of cells. (*e*) Watershed segmentation of the image in (*d*) using distance transformation. (*f*) Outline of the watershed-segmented cells in (*d*).

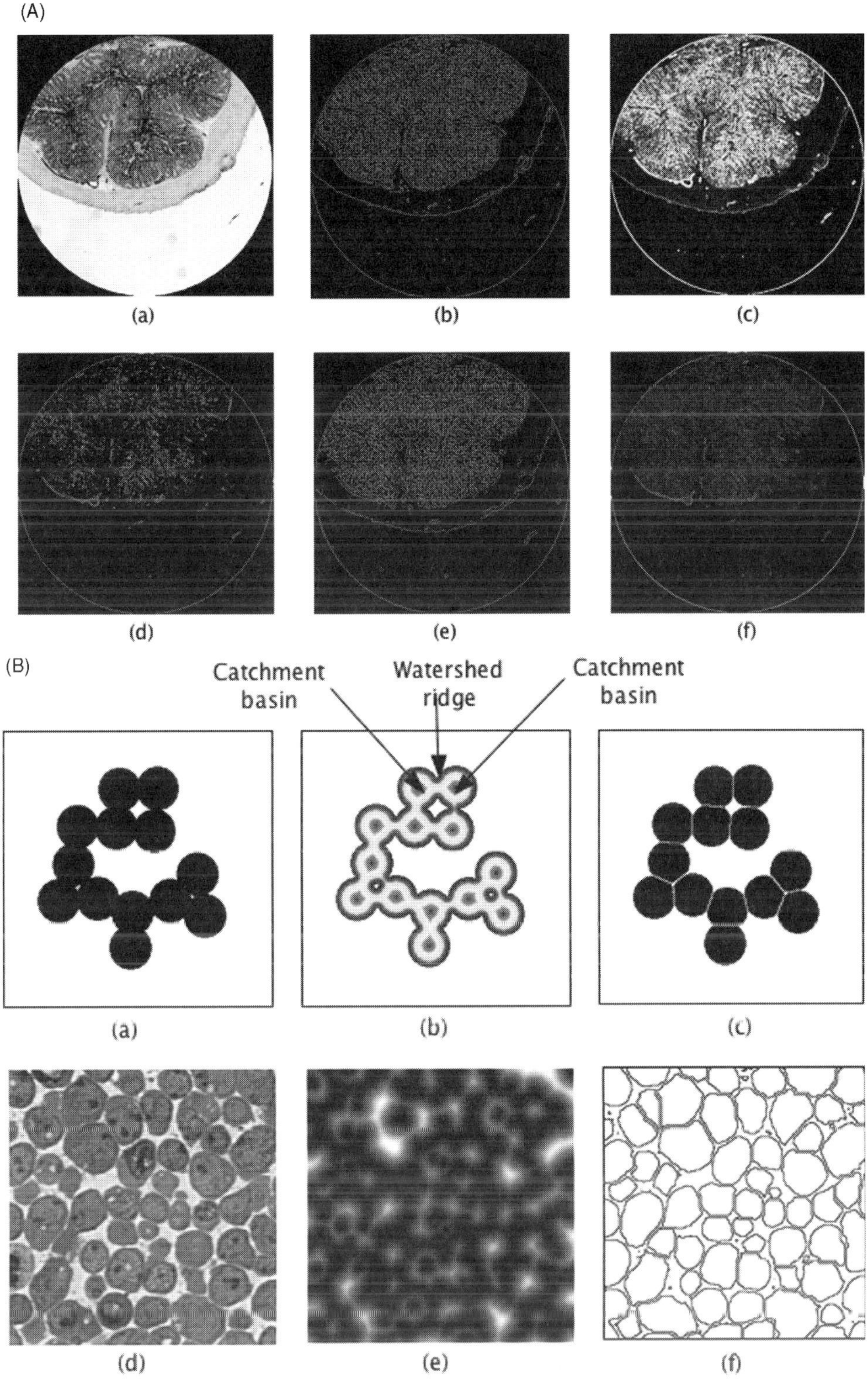
(A)
(a)
(b)
(c)
(d)
(e)
(f)
(B)
Catchment basin
Watershed ridge
Catchment basin
(a)
(b)
(c)
(d)
(e)
(f)

result, and a combination of methods may be more appropriate. Region-growing techniques are generally better in noisy images, where borders are extremely difficult to detect. An important property of regions is homogeneity and this is the main segmentation criterion in region growing where the basic idea is to divide an image into zones of maximum homogeneity. The criteria for homogeneity can be based on gray level, color, texture, or shape model using semantic information, etc. (14, 15). In the region-growing approach, a seed region is first selected and then expanded to include all homogeneous neighbors, and this process is repeated until all pixels in the image are classified. This technique works best on images with an obvious homogeneity criterion and tends to be less sensitive to noise because homogeneity is typically determined statistically (21).

Watershed segmentation

Another method for identifying different regions in an image is watershed segmentation (13). This is best understood by interpreting the intensity image as a landscape in which holes, representing minima in the landscape, are gradually filled in by submerging in water. As the water starts to fill the holes, this creates catchment basins and, as the water rises, water from neighboring catchment basins will meet. At every point where two catchment basins meet, a dam, or watershed, is built. These watersheds represent the segmentation of the image. This is a powerful tool for separating touching convex shapes. Watershed segmentation can be also implemented with sorted pixel lists, suggesting that segmentation can be performed very rapidly. (22). *Fig. 4(B)* (also available in the color section) illustrates the use of the watershed algorithm.

Texture-based segmentation

So far we have discussed segmentation methods based on image intensity; however, many images contain areas that are clearly differentiated by texture that could be used as a means of achieving

Figure 5. Image segmentation using texture-based segmentation, morphological operation, and color segmentation (see page xxii for color version).
(*A*) Texture-based segmentation. (*a*) Original image of a mouse kidney section. (*b*) Identification of kidney tissue from background (blue) and glomeruli identification (red). (*c*). Segmentation of cortex (green) and medulla (red). Note that these images were processed by CELLENGER software (images courtesy of The Wellcome Trust Sanger Institute/Definiens AG). (*B*) Removal of contamination by a morphological operation. (*a*) A TMA core image with a contamination spot (arrow). Note that the color blue represents the staining. (*b*) The mask image of segmentation for the staining without any morphological operations (red). (*c*) The mask image of segmentation for the staining after several morphological operations (green). Note that the contamination has been removed successfully in this mask (i.e. the spot is not green). (*C*) Common color segmentation approaches.

(A)

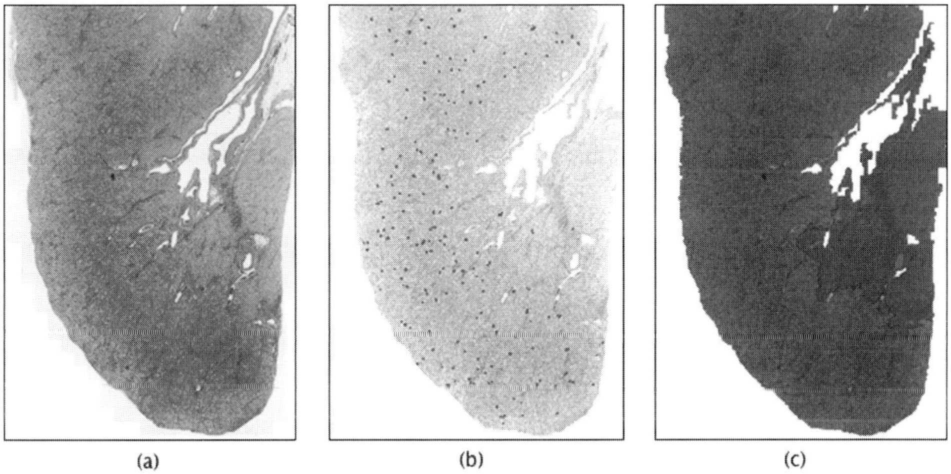
(a)
(b)
(c)

(B)

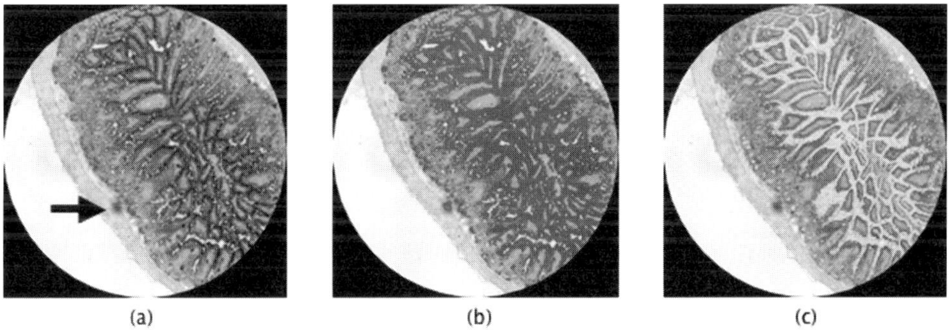
(a)
(b)
(c)

(C)

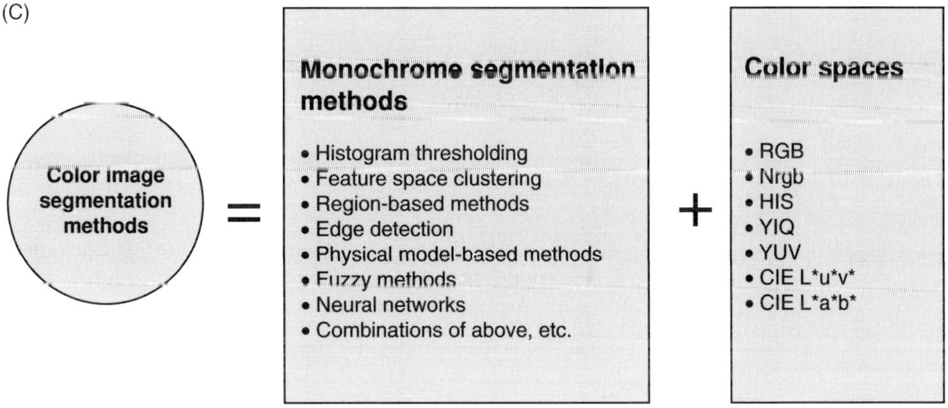
Color image segmentation methods
=
Monochrome segmentation methods
• Histogram thresholding
• Feature space clustering
• Region-based methods
• Edge detection
• Physical model-based methods
• Fuzzy methods
• Neural networks
• Combinations of above, etc.
+
Color spaces
• RGB
• Nrgb
• HIS
• YIQ
• YUV
• CIE L*u*v*
• CIE L*a*b*

segmentation. For example, in the kidney, the cortex and medulla can be differentiated from each other by the density and location of structures such as glomeruli (*Fig. 5A*, also available in the color section). Texture is characterized not only by the gray value at a given pixel, but also by the pattern in a neighborhood that surrounds the pixel (23). Texture features and texture analysis methods can be loosely divided into statistical and structural methods (24) where the following approaches can be applied: Hurst coefficient, gray level co-occurrency matrices, the power spectrum method of Fourier texture descriptors, Gabor filters, and Markov random fields (17, 23, 25). An example of texture-based segmentation is given in *Fig. 5(A)*.

Morphological operators

Contamination can be a source of problems in immunohistochemical staining. For example, dust particles or tissue fragments may break away and land on top of the region of interest, appearing as spots on the scanned image. Imperfections in slide surface chemistry or the immunohistochemical staining method can also potentially introduce artifacts. An example of contamination in a TMA core image is shown in *Fig. 5(B)* (also available in the color section). Such spots can be seen distributed in the slide background and on the surface of the tissue core. These defects have the ability to generate false results in signal quantification. Image-processing methods can be used to provide the best estimate of the true signal level by identifying the contamination and removing it. Various mathematical morphological operators can be used to identify the bona fide signal from background and contamination (12). These approaches combine both spatial and intensity information to distinguish true signal pixels from the background, with the contamination being trimmed off as outliers. It should be noted that careful specimen preparation will yield better images than any image-processing technique applied to images from a poor preparation.

Color image segmentation

Human eyes can discern thousands of color shades and intensities but only two-dozen shades of gray. Thus, objects that cannot be extracted using gray-scale values can often be extracted using color information. Compared with gray scale, color provides information in addition to intensity. However, the literature on color image segmentation is not as extensive as that of monochrome image segmentation. Most published results of color image segmentation are actually based on gray-level image-segmentation approaches with different color representations (26).

Generally, there are no standard rules to segment color images. *Fig. 5*(*C*) (also available in the color section) describes commonly used color-segmentation approaches.

2.2.4 Image registration and stitching

Image registration is the alignment of two or more images so that they are superimposed. This can be useful in comparing the precise co-localization of two different labels that have been imaged using different filters, or in comparing the same sample over time. Image stitching is when multiple frames are taken of a large tissue section and then recombined to create a composite image. Adjacent frames will overlap by a small amount and further mathematical processing can create seamless joins (e.g. the Multiple Image Alignment tool in the ANALYSIS package automatically generates seamless composite images). Both of these techniques are quite challenging, even when the images are identical or very similar. Accurately aligning images that are only moderately similar (e.g. different staining patterns on adjacent serial sections) presents a significant challenge to image registration algorithms, so the task is often aided by human intervention with the use of embedded markers for reference.

2.2.5 Image quantification

After segmentation and thresholding, an image is ready for image quantification. It should be recognized that, although the imaging process produces a high-resolution image for human viewing, one must be careful not to derive estimates of quantification by visual inspection alone. For example, quantification of immunohistochemical staining is usually performed subjectively according to a semi-quantitative grading, which can have poor reproducibility and depends on the skills and mood of the interpreter. Automatic image analysis has the potential to generate more robust and objective results than human viewing. It also provides a continuous scale for measurement and a statistical basis for developing classification criteria. This could have important implications in diagnosis, as accurate quantification of, for example, a tumor marker could help stratify patients more carefully into appropriate treatment regimes (27, 28). In TMA analysis, one of the required outputs is an estimate of the level of protein expression in a tissue core. *Fig. 6*(*a*, *b*) (also available in the color section) shows a pseudo-color intensity display that takes a gray-level image and maps the intensity range to a path in a color cube. *Fig. 6*(*c*) (also available in the color section) shows an analysis in which the percentage area and density of staining is computed.

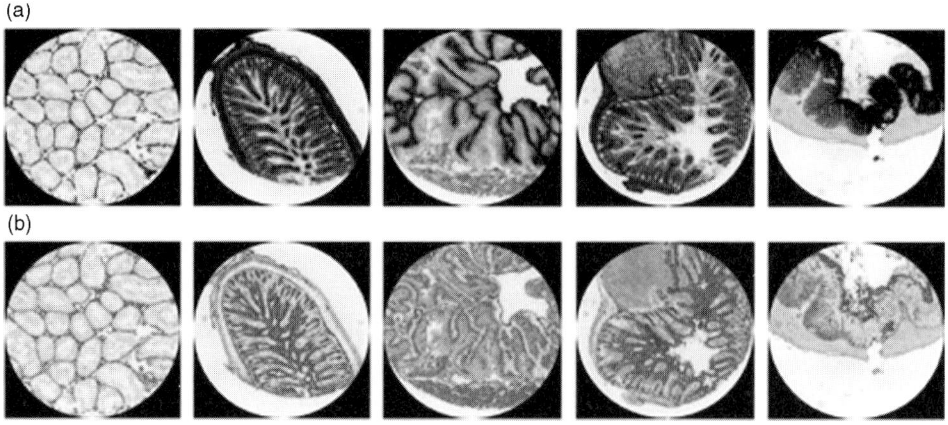

Tissue factor	0.785	0.600	0.836	0.777	0.498
Stain factor	0.156	0.898	0.433	0.718	0.556
Density factor	0.202	0.269	0.248	0.242	0.333

Figure 6. Example of staining intensity quantification (see page xxiii for color version). (*a*) Five different TMA cores with positive alkaline phosphatase staining (blue). (*b*) Segmentation of the blue staining in (*a*). Here, a pseudo-color intensity map has been derived from only the regions of blue staining and is shown laid over the TMA core; moderate/strong intensity is seen as yellow. (*c*) Quantitative analysis of the images in (*a*) by area analysis. The tissue factor is the tissue area/core area and represents the amount of tissue in the core. The stain factor is stain area/tissue area and represents the area of tissue stained. The density factor is the average stain intensity per pixel (normalized, where 0 is minimum and 1 is maximum intensity) and is a measure of signal strength.

2.2.6 Classification and clustering

After image quantification data have been extracted from an image, it is straightforward to carry out data classification or clustering. This is the process of finding models that separate two or more data classes on the data across a set of images (e.g. how many TMA cores show nuclear staining above a certain value). Multiple features can be accommodated into the analysis to derive complex relationships. Methods for clustering and classification analysis have been well used in pattern recognition and computational genomics and include decision tree induction methods, *k* nearest neighbors (*k*-NN), Bayesian methods, hidden Markov models (HMM), artificial neural networks (ANN), support vector machines (SVM), and prediction by collective likelihood of emerging patterns (29, 30).

2.2.7 Object-oriented image analysis

The challenge of understanding images is not just to analyze a piece of information locally, such as intensity, but also to bring the context into play. This suggests that methods other than looking at individual pixels would be of value. Recent developments in object-oriented image analysis afford much potential and extract complex morphological information from images (e.g. the CELLENGER package by Definiens AG; 31). Similar to human cognition principles, object-oriented approaches are based on the classification of image properties into semantic categories, with these categories being further processed and combined until the appropriate objects are identified. Importantly, such analyses combine various object properties together with their relationship to other objects. This is driven by a set of rules (i.e. algorithms and processing steps) that must be derived for each object under investigation. This process creates an image object hierarchy that describes different levels of objects from individual pixels to larger structures and their interrelationships. Once objects are identified, relevant statistical data can be extracted by subjecting them to morphometric analysis. An example of object-oriented image analysis is given in *Fig. 7* (also available in the color section) in which a section of a mouse embryo has been processed to identify various tissue types such as heart, kidney, and liver. Further rule sets could then be applied to determine signal strength and location in immunohistochemical staining

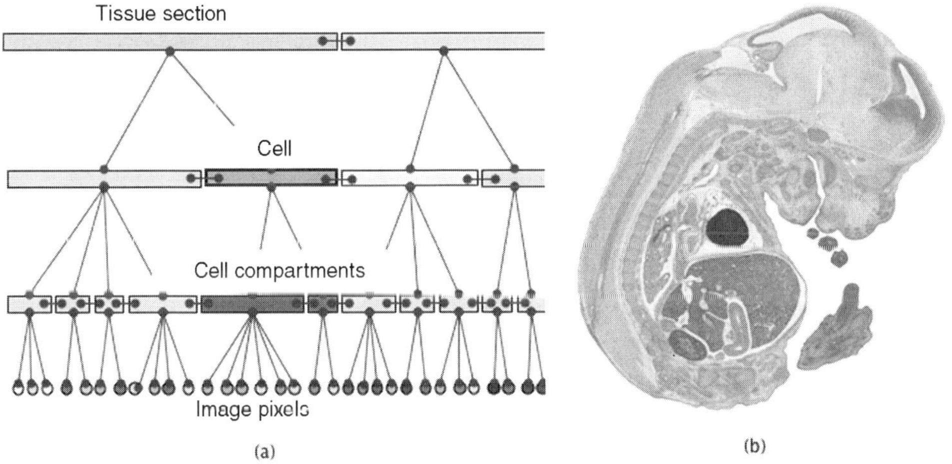

Figure 7. Object-orientated image analysis (see page xxiv for color version).
In this example, an E14.5 mouse embryo section was processed using a rule set in the CELLENGER package to identify uniquely different embryonic tissues. (*a*) Example of the image object hierarchies. (*b*) Result of image processing showing the separation of different tissue regions, e.g. heart (red); liver (yellow) and kidney (green) (images courtesy of The Wellcome Trust Sanger Institute/Definiens AG).

experiments on such sections. These approaches have many advantages including the ability to analyze low-contrast images or heterogeneous tissues samples. The only limitation in these approaches is the development of the relevant rules. However, packages such as CELLENGER by Definiens AG use a graphical meta-language that significant speeds up the process. The particular challenge of high-throughput TMA analysis is well suited to such approaches.

2.3 Image data handling

Given the enormous increase in immunohistochemical staining image data, it is clear that computational approaches, in combination with empirical methods, are expected to become essential for deriving and evaluating new hypotheses in biomedicine. To explore these valuable data fully, advanced bioinformatics infrastructures must be put in place to handle the volume, complexity, and dynamic nature of the image data, which are often collected and maintained in heterogeneous and distributed sources. A TMA experiment is a typical example of high-throughput research demanding reliable automated image processing to accelerate the discovery workflow and to reduce human error. Although several commercial and academic software tools are available, the majority have had limited success in handling the full spectrum of different types of TMA images. Furthermore, it is desirable that image data from high-throughput experiments should be integrated with other relevant databases to provide a common discovery platform for scientists. This remains a significant problem, and we are still some way from being able to perform fully integrated queries across multiple databases.

2.3.1 Image compression

Image processing is often very difficult, requiring significant computing power because of the large amounts of data used to represent an image. As technology moves forward, images will only increase in resolution; thus, there is a need to place sensible limits on the resulting data volume (15). Consider an example of a sagittal section through an E14.5 mouse embryo, where file size is a serious issue. If the whole section is imaged in color at 20× magnification with a large format 2k × 2k camera, it will require at least 120 frames to capture the whole section. At 12 MB per frame, this gives a file size in excess of 1.4 gigabytes for just one section. Given that several sections can be placed on one slide, it is easy to see that the amount of storage media required for archiving is enormous. One possible approach to decreasing the storage required is to work with compressed image data. For more detailed surveys of image-compression techniques, see (3), (32), and (33).

In choosing to compress image data, one should consider whether lossless or lossy compression is required. The advantage of lossy compression is that very high compression ratios can be reached in immunohistochemical staining images – in some cases reductions in file sizes of up to 80–90% can be achieved without perceptible effects on the visual image. Currently, JPEG (Joint Photographic Experts Group) is the most common form of lossy compression used, but one must consider carefully whether introduction of artifacts or distortions into highly compressed images will damage subsequent image-processing steps. Clearly, with lossless compression algorithms (such as TIF, LZW-TIF, and JPEG2000), the original image can be reconstructed but compression levels are more modest, perhaps up to 50% of the original file size.

Whilst JPEG was originally created as a standard in the late 1980s with a number of compression modes (34), only the baseline lossy mode has been popular. In today's imaging world, there is an ever-greater need for image compression, flexibility, and interchangeability. To meet these demands, a new standard with a wide set of features has been introduced, JPEG2000 (www.jpeg.org). JPEG2000 is a wavelet-based image-coding system that provides superior image quality at high compression ratios. The wealth of features provided by the JPEG2000 standard include lossy and lossless compression, embedding of metadata, support for large images, support for alternate color spaces, and progressive transmission by pixel accuracy. This latter point means that images can be reconstructed at different resolutions as needed for different target devices (e.g. only small images need be served over the web). Whilst decoding JPEG2000 images is efficient, encoding large images with JPEG2000 can be computer intensive. Also, plug-ins for the display of JPEG2000 images in web browsers are not yet universally available. Even so, any large imaging product should seriously explore the benefits of JPEG2000 for handling image compression.

2.3.2 Database structure and data integration

It is not unusual for a single high-throughput laboratory to generate in excess of 10 gigabytes of data per day. This amount of data presents two kinds of problem. First, there is a simple problem of storing and archiving the data. Secondly, the appropriate software tools are required to view, store, and query these data. Recently, a unified image structure has been proposed that considers time, space, and spectral bands as integral parts of the image (35). Such data structures have the potential to simplify the implementation of image databases and the subsequent analysis of the data contained within the database.

Metadata

Metadata is broadly defined as data about data, or data used to describe data, and includes two types of information: (i) technical metadata such as the height, width, color depth, file, and compression format of an image; and (ii) content metadata generated by imaging equipment and software, which includes: identification data such as date, laboratory, experiment, and experimenter; and annotations such as text and graphics that are added to the image based on their observation of the image (e.g. location of signal, pathology, etc.). Image analysis data such as image segmentation masks can be derived from an image by applying analysis tools to the metadata.

The maintenance of the correct linkage between images and their annotation can be a problem in a flat-file-based system. One method of maintaining data integrity is to embed annotation of the image contents directly into the image file headers, or embed the image file within a 'wrapper' file that can store this information (e.g. DICOM). Another method is to use Extensible Markup Language (XML) to maintain a linkage between flat files containing images and annotation. XML can be viewed as a powerful data model and a useful data exchange format, providing directly for a general data integration solution for biomedicine (36). This allows a flexible definition of a hierarchically nested set of tags and provides a number of advantages in managing the metadata. This makes images easier to share, search, and interpret by maintaining the association between the image and the metadata.

Another method of maintaining the linkage between images and their annotation is to use a relational database system (see below) to link the images and metadata together in a controlled fashion. This approach has the advantage that all of the relevant backup and security database tools are available for both images and metadata.

Database structure

Many different file and database management systems are currently in use with biological image data sets (37). Many widely used biological databanks hold data sets of flat files where different types of data are held in different files, e.g. one for images, one for annotation, etc. In common use, this has potential dangers in breaking the links between different parts of the data. For example, annotations can become disassociated from their correct image assignment.

An alternative database structure is the relational database (e.g. Oracle, MySQL). Relational databases describe all of the entities in the database through tables, as defined relations. Each row in a table is a record and each field in a row is an attribute. The relationships through these tables are based on Boolean algebra. The structure of relational databases can be well suited to biological data, as it permits the use of sophisticated interrogation and analysis tools. A key advantage of well-designed relational databases is that metadata is always linked to the primary image data and it can be hard to break this link inadvertently. One example of a relational image database is the Open Microscopy Environment (38), which is an open-source project that seeks to develop a relational database specification for microscopy (www.openmicroscopy.org). Finally, it is worth mentioning that other database structures exist, such as hierarchical and object-oriented architectures, that may be useful for handling image data. However, the query language implemented through such databases may be less structured than in the relational database.

A database system that aims to be a general integration mechanism for all of the data generated in the high-throughput environment should satisfy at least the following conditions (39): (i) it must be able to compile any queries to extract relevant information from the database based solely on the structure of the query and not on the availability of the database schema; (ii) it should use a data model and exchange format that external databases and other software systems can easily translate; (iii) it must shield existing queries from evolution of the external sources as much as possible. For example, an extra field appearing in an external database table must not necessitate the recompilation or rewriting of existing queries over that data source.

Structured query language (SQL)

There is a need in image databases to have a standard tool to query information contained within that database. Many query languages developed for image information systems are command languages, or commands plus expressions. As the size of the image database grows, efficient querying routines are required to retrieve relevant data concerning the image and its properties. Most of the software packages provided with hardware include simple querying routines. Whilst simple search routines or filter systems are based on simple Boolean operations, more advanced systems utilize the full range of SQL tools. An advantage of using an SQL-like syntax is that it facilities the export and transfer of data into other databases. For more details on SQL, see the tutorial available at www.w3schools.com/sql/default.asp.

Acknowledgements

We are most grateful to Gareth Maslen and Stephen Keenan for critical reading of this manuscript. We also thank all members of Teams 86 and 39 at the Sanger Institute for their support. This work is funded by the Wellcome Trust.

3. REFERENCES

1. **Ramdas L, Coombes KR, Baggerly K, *et al.*** (2001) *Genome Biol.* **2**, research0047.1-research0047.7.
2. **Feng Z-P** (2002) *In Silico Biol.* **2**, 291–303.
3. **Gonzalez RC & Woods RE** (2002) *Digital Image Processing*, 2nd edn. Addison-Wesley, MA, USA.

★★ 4. **Kallioniemi OP, Wagner U, Kononen J & Sauter G** (2001) *Hum Mol Genet.* **10**, 657–662. – *Original review on TMA technology.*

★★ 5. **Warford A** (2004) *Expert Rev. Proteomics*, **1**, 283–292. – *Recent survey of TMA approaches and their significance.*

6. **Simon R, Mirlacher M & Sauter G** (2003) *Expert Rev. Mol. Diagn.* **3**, 421–430.

★★★ 7. **Sharpe J, Ahlgren U, Perry P, *et al.*** (2002) *Science*, **296**, 541–545. – *Illustrating the exciting new technology of optical projection tomography (OPT).*

8. **Kamberova G & Shah S** (2002) *DNA Array Image Analysis: Nuts & Bolts.* DNA Press, Skippack, PA, USA.
9. **Jain AN, Tokuyasu TA, Snijders AM, Segraves R, Albertson DG, & Pinkel D** (2002) *Genome Res.* **12**, 325–332.

★★ 10. **Yang YH, Buckley MJ, Dudoit S & Speed TP** (2002) *J. Comput. Graph. Stat.* **11**, 108–136. – *Comparison of methods used in image analysis on cDNA microarray data.*

11. **Daubechies I** (1992) *Ten Lectures on Wavelets.* Society for Industrial and Applied Mathematics, Philadelphia, PA, USA.
12. **Serra J (ed.)** (1988) *Image Analysis and Mathematical Morphology*, 2nd edn. Academic Press, NY, USA.
13. **Soille P** (2003) *Morphological Image Analysis – Principles and Applications*, 2nd edn. Springer-Verlag, NY, USA.

★★★ 14. **Russ JC** (2002) *The Image Processing Handbook*, 4th edition. CRC Press, Boca Raton, FL, USA. – *A comprehensive digital image processing textbook.*

15. **Castleman KR** (1996) *Digital Image Processing*, Prentice Hall, NJ, USA.
16. **Skarbek W & Koschan A** (1994) *Colour Image Segmentation: a Survey.* Technischer Bericht 94-32, Technical University of Berlin, Germany.
17. **Watt A & Policarpo F** (1998) *The Computer Image.* Addison-Wesley, NY, USA.

★★ 18. **Sahoo PK, Soltani S & Wong AKC** (1998) *Comp. Vision Graph. Image Process.* **41**, 233–260. – *A survey of thresholding techniques for image segmentation.*

19. **Sezgin M & Sankur B** (2004) *J. Electron. Imaging* **13**, 146–165.
20. **Otsu N** (1979) *IEEE Trans. Syst. Man Cybern.* **9**, 62–66.

★ 21. **Adams R & Bischof L** (1994) *IEEE Trans. Pattern Anal. Mach. Intell.* **16**, 641–647. – *Seeded region growing for image segmentation.*

★ 22. **Vincent L & Soille P** (1991) *IEEE Trans. Pattern Anal. Mach. Intell.* **13**, 583–597. – *Watershed for image segmentation.*

23. **Wu J** (2002) *Rotation Invariant Classification of 3D Surface Texture using Photometric Stereo.* PhD thesis, Heriot-Watt University, Edinburgh, UK.

★ 24. **Haralick RM** (1979) *Proc. IEEE*, **67**, 786–804. – *Extracting statistical and structural texture features.*

★★ 25. **Reed TR & du Buf JMH** (1993) *Comp. Vision Graph. Image Process.* **57**, 359–372. – *A review of texture segmentation and feature extraction techniques.*

26. **Chen CH, Pau LF & Wang PSP (eds.)** (1993) *Handbook of Pattern Recognition and Computer Vision.* World Scientific, Singapore.

★★★ 27. **Camp RL, Chung G & Rimm DL** (2002) *Nat. Med.* **8**, 1323–1328. – *Development of automated image analysis for TMAs.*

★★ 28. **Rhodes A, Borthwick D, Sykes R, Al-Sam S & Paradiso A** (2004) *Am. J. Clin. Pathol.* **122**, 51–60. – *Long-term study on the standardization of immunohistochemistry by image analysis.*

29. **Ripley BD** (1996) *Pattern Recognition and Neural Networks.* Cambridge University Press, Cambridge, UK.

30. **Grant RP (ed.)** (2004) *Computational Genomics.* Horizon Bioscience, Wymondham, UK.

31. **Biberthaler P, Athelogou M, Langer S, Luchting B, Leiderer R & Messmer K** (2003) *Eur. J. Med. Res.* **8**, 275–282.

32. **Rabbani M & Jones PW** (1991) *Digital Image Compression Techniques.* SPIE Optical Engineering Press, Bellingham, WA, USA.

33. **Clarke RJ** (1995) *Digital Compression of Still Images and Video.* Academic Press, London/New York.

34. **Acharya T & Tsai P-S** (2004) *JPEG 2000 Standard for Image Compression: Concepts, Algorithms and VLSI Architectures.* John Wiley & Sons, Chichester, UK.

35. **Andrew PD, Harper IS & Swedlow JR** (2002) *Traffic,* **3**, 29–36.

36. **Achard F, Vaysseix G & Barillot E** (2001) *Bioinformatics,* **17**, 115–125.

37. **Lesk AM (ed.)** (2005) *Database Annotation in Molecular Biology – Principles and Practice.* John Wiley & Sons, Chichester, UK.

★★ 38. **Swedlow JR, Goldberg I, Brauner E & Sorger P** (2003) *Science,* **300**, 100–102. – *A description of the Open Microscopy Environment (OME).*

39. **L. Wong** (2000) *J. Funct. Program.* **10**, 19–56.

CHAPTER 9

Quality assurance in immunochemistry

Peter Jackson

1. INTRODUCTION

Quality control (often referred to as internal quality control) is a part of quality assurance. It comprises all of the different measures taken to ensure the reliability of investigations and is not restricted to technical procedures. Quality control is carried out to minimize variability and to guarantee the quality of laboratory results.

In the laboratory, it can be defined as the analysis of material of known content in order to determine in real time whether the procedures are performing within predetermined specifications. Examples of this are the use of known positive and negative control sections, which are stained alongside the test sections. If the control sections do not demonstrate the expected staining pattern, the test sections are not reported. In this way, quality control controls the release of reports/results.

Quality control differs from quality assessment in that it controls the issue of results in real time, on a daily basis. Quality assessment is the challenge of quality control procedures by specimens of known but undisclosed content. It is often done by taking part in external quality assessment schemes (see section 3). It is a retrospective activity that does not affect the day-to-day issue of results. External quality assessment does establish 'between laboratory' comparability, which in turn allows the identification of best practice and poor performance.

Standardization of the basic procedures used in histopathology such as fixation, tissue processing, and the treatment of tissue sections prior to staining should make a major contribution in improving laboratory performance. However, tissue processing is far from standardized. The formulation of the fixative used has traditionally been left to each

Immunohistochemistry: *Methods Express* (S. Renshaw, ed.)

individual laboratory and similarly tissue processing to paraffin wax is also far from standardized. Laboratories select and employ dehydrating agents of their choice. Tissue processing times vary considerably from laboratory to laboratory with most laboratories having adopted rapid or extended processing times to suit individual tissue types.

This lack of standardization of the fundamental processes of histology culminates in the production of a unique preparation. Subsequently, sections cut from such blocks provide substrates for both tinctorial and immunochemical investigations that may be vastly dissimilar to sections cut from material fixed, processed, and sectioned elsewhere. In dealing with such material, it is essential that laboratory personnel understand the problems and pitfalls associated with the basic steps in the preparation of tissue sections for histological investigations and they must be fully aware of the established ways of manipulating protocols in order to give high-quality immunochemical staining on their own and referred sections.

Successful immunochemistry may be seen as the correct integration of several technical parameters. The aim of immunochemistry is to achieve reproducible and consistent demonstration of antigens with the minimum background staining whilst preserving the integrity of tissue architecture.

2. METHODS AND APPROACHES

The following are essential considerations for immunochemistry and will each be considered in turn:

- Fixation and tissue processing
- Microtomy
- Decalcification
- Antigen-retrieval techniques
- Immunochemical methodologies
- Controls
- Microscopic interpretation
- Background staining
- Troubleshooting
- Quality assurance schemes

2.1 Fixation and tissue processing

Adequate and appropriate fixation is the cornerstone of all histological and immunochemical preparations. Tissue fixation should therefore be adequate and appropriate. Standard operating procedures regarding fixation should be rigidly adhered to. Good fixation is the delicate balance between underfixation and overfixation. Ideal fixation is the balance between good morphology and good antigenicity. A diagnosis is made on morphological clues and pattern recognition. If the section is of poor quality – and this is usually associated with poor fixation – then diagnosis can be unnecessarily difficult and/or incorrect, leading to inappropriate treatment or no treatment.

The demonstration of many antigens depends heavily on the fixative employed and on the immunochemical method selected. Poor fixation, or delay in fixation, causes loss of antigenicity or diffusion of antigens into the surrounding tissue. It is unfortunate that no one fixative is ideal for the demonstration of all antigens and some tissue antigens necessitate the use of frozen sections.

In order to achieve the above, a fixative must obtain rapid penetration into the tissue. This is more appropriate when considering formaldehyde, as this is most commonly used on pieces of tissue prior to processing to paraffin wax. Precipitant fixatives tend to be employed on pre-cut tissue sections or cytological preparations.

2.1.1 Cross-linking fixatives

To ensure rapid penetration of the tissue by any fixative (and hence, rapid fixation), the smallest possible pieces of tissue must be taken from the gross specimen. Although sample size must be taken into consideration, tissue pieces should be no bigger than the tissue cassettes that they are processed in and should fit into the cassettes without being crushed by the lid. Crushing of tissues can cause nonspecific immunochemical staining patterns. For gross specimens, tissue penetration can be enhanced further by perfusion or by cutting into the specimen before it is submerged in fixative.

Standardization of fixation in the USA has been addressed by the College of American Pathologists. Their recommendations are that, for optimal results, tissue specimens with maximum dimensions of 1.0 × 1.0 × 0.4 cm should be fixed for a minimum of 3.5 h and a maximum of 18 h in fresh neutral buffered formaldehyde (formalin). The neutral buffered formaldehyde must be less than 1 month old. Formaldehyde-fixed tissues that cannot be processed immediately for paraffin embedding should be stored in 70% alcohol until processed.

Tissue penetration by formaldehyde is rapid due to its low molecular weight, with initial binding to proteins occurring within 24 h (1). Although methylene bridge formation can take several days to reach an optimum, with immunochemistry this is not considered so much of a problem. Optimal fixation leads to a high degree of antigen masking, requiring antigen retrieval in order to allow antibody binding. As a result, 24 h is generally considered to be an optimal fixation time, providing a good compromise between obtaining good cytological preservation and immunolocalization, whilst keeping antigen masking to a minimum (2). Tissue intended for processing to paraffin wax should be stored long term in 70% alcohol in order to maintain antigenicity, following the initial 24 h formaldehyde fixation. The same applies to tissues sitting in automated processors over weekends (2).

Poorly fixed tissue blocks do not process to paraffin wax adequately. Alcohol used in tissue-processing dehydration steps is an excellent fixative and an excellent dehydrating agent. However, if asked to perform both as a fixative and as a dehydrant, as with poorly fixed tissue blocks, it fails to achieve either and a poorly processed paraffin block ensues. If poor processing occurs regularly within a laboratory, it pays to re-examine the fixation protocols rather than placing the blame solely on the quality of the alcohol. However, in some instances reprocessing of poorly processed blocks may be carried out. It may be necessary only to reinfiltrate with paraffin wax, to reprocess to xylene, or to reprocess to alcohol. In all cases, reasonable immunochemical staining can be achieved for most antigens.

2.1.2 Precipitant fixatives

Alcoholic fixatives, usually methanol, acetone, or combinations of these, used either alone or in combination with formaldehyde, are often used for whole-cell preparations or for cryostat sections using fresh frozen tissue. However, frozen sections have certain inherent disadvantages when compared with their paraffin counterparts. These include:

- Poor morphology in comparison with their paraffin counterparts.
- Poor resolution at higher magnifications.
- Availability of fresh tissue. Special arrangements are required for the collection and storage of fresh unfixed tissues.
- Limited retrospective studies.
- Hazards to health associated with the handling of fresh unfixed material, e.g. human immunodeficiency virus, hepatitis B virus, and tuberculosis.
- Receiving tissue in the laboratory as soon as possible after removal. Good liaison with clinicians and theatre staff is essential.

Alcohol (and acetone) penetrates tissue poorly and is generally only used on tissue sections or cytological preparations as opposed to pieces of tissue. However, some laboratories do utilize alcohol on small pieces of tissue and obtain good cytological definition. Alcohol is often employed in other fixative mixtures to reduce the time spent during tissue processing, for example Carnoy's (methacarn) fixative for demonstrating nucleic acids.

Short acetone fixation alone does not completely stabilize frozen sections against the detrimental effects of long incubations in aqueous solutions. Following long immunochemical staining techniques, acetone-fixed frozen sections may show morphological changes such as chromatolysis and apparent loss of nuclear membranes. It has been observed that such changes to tissue morphology can be prevented by ensuring that the tissue sections are thoroughly dried both before and after fixation in acetone. Treating fresh tissue as described in *Protocol 1* will generally produce high-quality immunochemical staining.

Protocol 1

Preparation of frozen sections for immunohistochemical staining

Equipment and Reagents

- Pre-cooled isopentane in liquid nitrogen
- 3-Aminopropyltriethoxysilane (APES)-coated slides or Super Frost Plus slides (Fisher Scientific)
- Acetone

Method

1. Snap freeze small pieces of tissue (5 × 3 × 3 mm) using pre-cooled isopentane in liquid nitrogen (see Chapter 4, section 2.2.2).
2. Cut frozen sections at 5 μm. Pick up on APES-coated or Super Frost Plus slides.
3. Air dry sections overnight at room temperature.
4. Fix sections in acetone (room temperature) for 10 min.
5. Air dry sections.
6. Proceed with immunochemical staining protocol or store sections at –20°C.

In view of the disadvantages associated with the production of frozen sections, the general trend is one away from the use of frozen sections and most immunochemical investigations performed in diagnostic histopathology laboratories are now carried out on formaldehyde-fixed, paraffin-embedded tissue sections.

2.2 Microtomy

It is essential that immunochemical stains are carried out on well-fixed, well-processed, high-quality paraffin sections. Sections should be of nominal thickness and free from the artifacts associated with microtomy. Air bubbles under tissue sections can trap antibodies and chromogen (see

Fig. 1). Knife scores enhance artifacts around the knife tract. Displaced tissue can obscure positively stained cells, and debris from the water bath may make interpretation difficult. Debris can usually be identified from the actual tissue section, as it appears in a slightly different focal plane. If sections are too thick, reagents can be trapped between cell layers. This also happens in areas where the tissue is folded or creased. Sections for immunochemistry should be picked up on APES-coated slides, poly-L-lysine-coated slides, or Super Frost Plus slides, otherwise they are unlikely to survive the techniques associated with heat-mediated antigen retrieval.

Slides should be dried overnight at 37°C or can be dried more rapidly in a 60°C oven. However, exposing tissue to 60°C can have a deleterious effect on the staining of some antigens, for example vimentin. The duration and intensity of heating tissues during embedding or slide drying should therefore be kept to a minimum and certainly no longer than overnight.

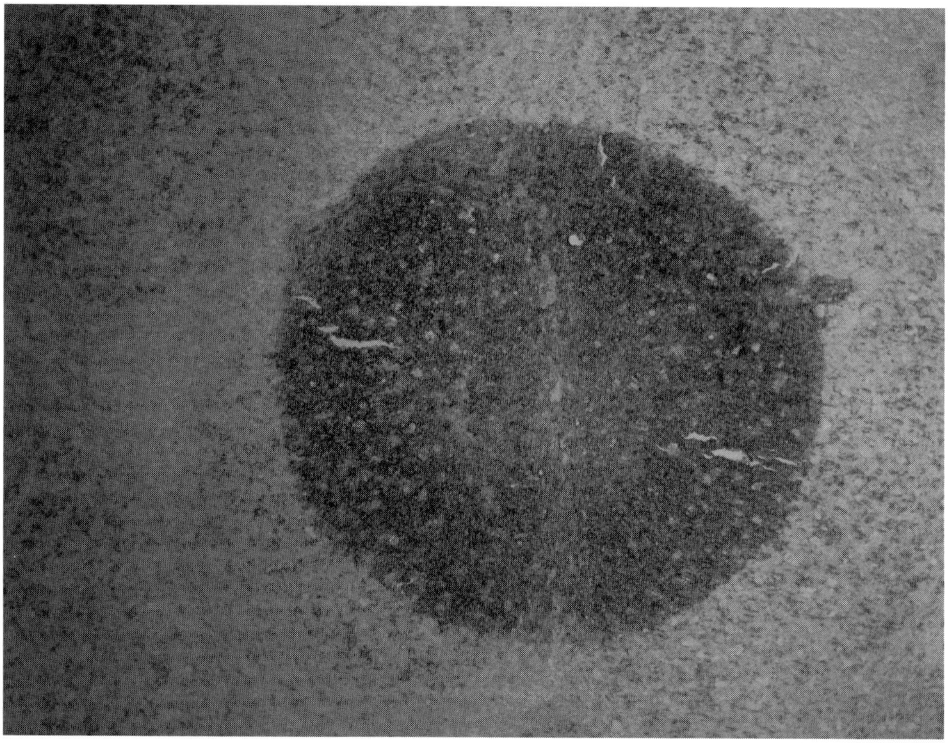

Figure 1. Example of a microtomy artifact.
An air bubble under the section has trapped antibodies and chromogen, leading to the observed nonspecific artifact.

2.3 Decalcification

It is essential that a specimen is adequately fixed prior to decalcification. Laboratories should choose a decalcifying reagent that has been evaluated in their own laboratory and has been shown not to destroy the antigen in question. This evaluation should be carried out using both tumor and normal tissues. CD15 is an example of a marker that may be destroyed on the tumor by decalcification but may remain detectable on normal cells. However, several researchers have reported that decalcification does not appear to be deleterious to any great degree on the demonstration of numerous antigens. Athanasou (3) observed that decalcification in strong acids did not diminish the reactivity of antigens such as LCA, S100, and EMA. Weaker acids (formic, acetic) and EDTA showed greater preservation of antigenicity with the added advantage of better morphology. TCA showed merit as a one-step fixation/decalcifying agent for both paraffin-embedded and frozen sections. Matthews (4) and Mukai (5) and co-workers demonstrated similar findings on other antigens.

Conversely, Miller (2) claims to have observed damaging effects of decalcification on antigens such as CD43, Ki67, ER, and PR, and discourages the use of strong acids as decalcifying agents. This further enforces the need for laboratories to choose a decalcifying reagent that has been evaluated in their own laboratory and has been shown not to destroy the antigen in question.

2.4 Antigen retrieval

In many instances, immunoreactivity can be restored without compromising the cellular morphology of the tissues.

2.4.1 Proteolytic (enzymatic) antigen retrieval

This can be accomplished through the use of a protease before immunochemical staining. The use of proteolytic enzyme digestion on formaldehyde-fixed, paraffin-embedded tissue sections was first described by Huang *et al.* (6). The enzyme-based retrieval method increased the range of useful antibodies that could be applied to routinely fixed, paraffin embedded tissue sections in diagnostic histopathology. Not all antigens require proteolytic digestion and care must be taken not to create 'false' antigenic sites or to alter existing antigenic sites. In some instances, immunochemical staining may be reduced or absent following enzyme digestion.

For general use, trypsin is probably the most widely used proteolytic enzyme used for enzymatic antigen retrieval. Care should be taken in the selection of the trypsin. Crude and relatively inexpensive trypsin containing the common impurity chymotrypsin often performs best. In fact, chymotrypsin is probably the active ingredient, as purified trypsin gives poor results and chymotrypsin alone can be used successfully (7). Other popular enzymes include pronase, proteinase K, and pepsin. A few antigens such as IgE are said to be preferentially revealed by protease XXIV.

The choice of protease, its concentration, and its duration of digestion are largely empirical. It is, however, important to optimize the use of the enzyme of choice, rather than to use a broad range of enzymes.

Key factors for enzyme digestion are:

- Temperature
- pH
- Enzyme concentration
- Duration of digestion

For optimum immunochemical staining, digestion time is critical and depends on the duration of fixation in formaldehyde. Tissues fixed in formaldehyde for long periods usually require prolonged exposure to proteolytic enzyme. It may be found that digestion time varies because of batch-to-batch variation in enzyme activity. Another pitfall with enzymatic antigen retrieval is that poorly fixed tissues are easily overdigested with a resulting loss of morphological detail.

Where fixation and paraffin-processing schedules are not known, as with referred blocks and slides, it may be necessary to perform a range of digestion times. In these cases, it is worth recording the optimum digestion time, which may be useful for future immunochemical staining studies on material sent from the same source.

For the above reasons, it may not be wise to recommend any one protease over another. Each laboratory must experiment with several proteolytic enzymes using control sections cut from tissue blocks fixed and paraffin processed in their own department in order to identify the optimal conditions for their own material.

In summary:

- Only aldehyde-fixed tissues require protease digestion.
- There are only a limited number of antigens that benefit from the procedure.
- The procedure must be optimized using the laboratory's own material.
- Failure to optimize the protease digestion time to the duration of fixation may produce false-negative results.

- The longer the duration of fixation, the longer the protease digestion time required.

2.4.2 Heat-induced epitope (antigen) retrieval

Heat-induced epitope (antigen) retrieval (HIER) has proved to be a revolutionary technique, as many antigens previously thought to have been lost or destroyed by fixation and paraffin processing can now be recovered and demonstrated. Antibodies such as MIB1 (Ki67) and estrogen receptor, which were previously only demonstrable in frozen tissue sections, now work well following heat pre-treatment on paraffin sections.

However, the theory behind HIER remains unclear. To date there are three hypotheses:

1. Heavy metal salts (protein-precipitating fixatives) act as secondary fixatives following primary tissue fixation in formaldehyde and protein-precipitating fixatives frequently display better preservation of antigens than cross linking aldehydes.
2. During formaldehyde fixation, inter- and intramolecular cross-linkages alter the protein conformation such that it may not be recognized by some antibodies. HIER removes the weaker Schiff bases but does not remove the methylene bridges. The resulting protein conformation is intermediate between fixed and unfixed.
3. Calcium coordination complexes formed during tissue fixation prevent antibodies from combining with epitopes on tissue-bound antigen by steric hindrance. High temperatures weaken or break some of the calcium coordinate bonds, but the effect is reversible on cooling and the complexes reanneal. The presence of a chelating agent at the particular temperature at which the coordinate bonds are disrupted removes the calcium complexes and hence the steric hindrance.

It is clear that the mechanism(s) of HIER is poorly understood and as yet there has been no convincing explanation regarding the rationale of the mechanisms involved. However, without the application of such techniques, there is no doubt that immunochemical staining is vastly inferior. This application illustrates the fact that histopathology laboratories often stand alone in the field of medical laboratory science and employ techniques about which little is known to improve staining in tissue sections.

The following are the commonly used methods of HIER:

1. Microwave oven heating. This involves the boiling of sections either in a Coplin jar or in a plastic microwaveable container. Whilst the microwave oven Coplin jar method allows good reproducible results for

many antigens, the limited numbers and the constant attention required to ensure that the sections do not dry out makes this method very time-consuming. The alternative large batch microwaving does suffer from inconsistencies, thought to be due to 'hot' and 'cold' spots creating an unbalanced retrieval of antigens, and the rigorous boiling of tissues can lead to dissociation from the slide. Scientific models have features such as stirrers and temperature monitoring probes to try and alleviate these problems. To help prevent section dissociation from the slide, always use APES-coated slides or similar.

2. Pressure cooking. This method does not tend to suffer from such inconsistencies and is far less time-consuming. Depending on the capacity of the pressure cooker, up to 75 slides can be retrieved at the same time. Pressure cooker antigen retrieval is carried out under full pressure, when the temperature of the retrieval solution reaches 120°C. The 20°C difference in temperature between pressure cooker and microwave retrieval appears to be beneficial for the demonstration of many antigens and is the retrieval method of choice in many diagnostic laboratories. The pressure cooker seal should be changed regularly. A defective seal leads to prolonged boiling. The time taken for the pressure cooker to reach maximum pressure for each individual run should be noted and when times begin to vary the seal should be changed.
3. Microwave pressure cooking. This uses a microwaveable pressure cooker, which is brought to full pressure by heating in the microwave oven. Such pressure cookers are relatively inexpensive and can be purchased from leading manufacturers of domestic plastics. Scientific microwave ovens can come equipped with purpose-built, on-board pressurized vessels.
4. Decloaker device. A decloaker is a recently introduced, digitally controlled electric pressure cooker into which up to four separate containers of antigen-retrieval solutions may be placed. As the heating cycle is controlled digitally, inconsistencies between batches are minimized.
5. Vegetable steamer. This method involves the use of a domestic vegetable or rice steamer. Antigen-retrieval solution is brought to near-boiling temperatures by internally generated steam. The major disadvantage of this method is the extended times of retrieval.
6. Combinations of HIER and proteolytic enzyme digestion. This involves the microwave method followed by a brief digestion in trypsin. This technique can be used for so-called 'difficult' antigens and is useful when dealing with bone marrow trephines and renal biopsies to give reliable light-chain demonstration of myeloma and AL amyloid.

There is a wide range of antigen-retrieval solutions available, which often makes selection difficult. In any laboratory it is important that retrieval solutions are evaluated using control sections produced by that laboratory. It may be found that three or four retrieval solutions will be selected for use as these may give optimal demonstration for certain antigens of interest to that laboratory. Popular antigen-retrieval solutions include citrate buffer (pH 6.0) and Tris/EDTA (pH 9.0). They can be divided into two groups:

1. Carboxylic and organic compounds used at pH 6.0:
   - Citric acid
   - Sodium carbonate
   - Sodium bicarbonate
   - Urea
   - Maleic acid
   - Sodium acetate
   - EDTA

2. Metal and salt solutions:
   - Aluminum chloride
   - Sodium chloride
   - Sodium fluoride
   - Zinc sulfate
   - Lead thiocyanate
   - Calcium chloride
   - Nickel chloride
   - Magnesium chloride
   - Ammonium chloride

There are a number of advantages and disadvantages of HIER. The main advantage is that heating times to retrieve antigens tend to be standard, regardless of the duration of fixation. This is in contrast to the variability in digestion times required when using proteolytic enzyme digestion. Others advantages include:

- Increased intensity of staining and the number of cells stained.
- Demonstration of antigens that are not normally demonstrable in formaldehyde-fixed, paraffin-embedded tissue.
- Production of consistent, reliable, high-quality immunochemical staining of formaldehyde-sensitive antigens.

The main disadvantage when performing HIER is that extreme care must be taken not to allow the sections to dry (so called 'flash drying'), as this destroys antigenicity and produces section artifacts. Antigen-retrieval solutions should be flushed from the container with cold running water. The sections can then be removed when the fluid is cool. Other disadvantages are associated with over-retrieval and include:

- Destruction of antigenicity giving false-negative staining.
- False-positive staining caused by binding other than antibody to antigen.
- The tissue can appear ragged.
- The tissue sections can become dissociated from the slides.
- The tissue may lose its ability to take up counterstain.
- Mucinous areas may be lost.
- Nonspecific staining may be increased.

However, the advantages of HIER far outweigh the disadvantages and therefore it is essential that the techniques involved in HIER are carried out correctly.

2.5 Immunochemical methodologies

There are numerous immunochemical staining techniques (see Chapter 4, section 2.10.2). The selection of a suitable technique should be based on parameters such as the type of preparation under investigation e.g. frozen section, paraffin, or cytological preparation, and the degree of sensitivity required. Many techniques, including the direct method, the indirect method, peroxidase anti-peroxidase, and alkaline phosphatase anti-alkaline phosphatase, have now been superseded by more sensitive methods. Good immunochemistry relies on good detection. Many antigens are depleted by the processes involved in fixation and paraffin processing and it is of the utmost importance to employ a modern sensitive detection and visualization system. Immunochemical staining based on the affinity of streptavidin and biotin (avidin–biotin complex, or ABC) has become the most popular in recent years and more recently the extended polymer-labeled antibody detection systems such as EnVision (Dako) and ImmPRESS (Vector) are gaining in popularity. The great advantage of these detection systems is that they are not based on the streptavidin–biotin detection methods and therefore endogenous biotin does not cause problems in interpretation. Such methods are at least as sensitive as the traditional ABC methods and have the added advantage that they are quicker and easier to perform. The extended labeled polymer methods are based on the two-layer indirect technique and utilize pre-diluted and ready-to-use enzyme-labeled second-layer antibodies.

2.5.1 Avidin–biotin techniques

These methods utilize the high affinity of the glycoprotein avidin for biotin (vitamin H). Biotin can be conjugated to a variety of biological molecules, including antibodies. One molecule of avidin can combine with four biotin molecules and this affinity is made use of in the ABC systems. However, avidin has two distinct disadvantages when used in immunochemical detection systems. It has a high isoelectric point of approximately 10 and is therefore positively charged at neutral pH when used in immunochemical staining methods. Consequently, it may bind nonspecifically to negatively charged structures such as the nucleus. The second disadvantage is that avidin is a glycoprotein and reacts with molecules such as lectins via the carbohydrate moiety. This reaction may be blocked by using an analogue of the carbohydrate on the avidin, e.g. 0.1 M α-methyl-D-mannoside can be added to the solution containing the avidin.

Both problems may be overcome with the substitution of streptavidin for avidin. Streptavidin is a protein isolated from the bacterium *Streptomyces avidini* and like avidin has four high-affinity binding sites for biotin. Streptavidin has an isoelectric point close to neutral pH and therefore will not bind to positively charged structures at the near-neutral pH used in immunochemical staining systems. Furthermore, streptavidin is not a glycoprotein and therefore does not bind to lectins. The physical properties of streptavidin make this protein much more desirable for use in immunochemical staining systems than avidin.

The more sophisticated and sensitive extensions of the streptavidin–biotin technique employ pre-formed complexes. Streptavidin and biotinylated enzyme are simply mixed together at appropriate concentrations and allowed to stand for 30 min to allow the complex to form. Failure to leave the complex for the required amount of time will lead to a weak signal. The pre-formed complex is then added to the biotinylated antibody. Careful stoichiometric control ensures that some binding sites remain free to bind with biotinylated antibody. This allows the pre-formed complex to bind and provides a very high signal at the antigen-binding site. Horseradish peroxidase or alkaline phosphatase can be used as the enzyme label.

As a word of warning, because pre-diluted secondary and tertiary antibodies are titered to work optimally in a specific kit, do not attempt to mix reagents from different kits.

2.5.2 Optimal dilution of primary antibody

The optimum dilution (titer) of primary antibody for immunochemistry is the concentration of the primary antibody that gives the optimal specific staining with the least amount of background staining.

The optimal dilution will depend on the type and duration of fixation. Serial dilutions of antibody will often give the distribution of reactivity shown in *Fig. 2*. The optimum concentration of primary antibody is that measured below the apex of the peak and the use of several sections with varying densities of antigen will aid the determination of a correct working dilution.

2.6 Controls

The importance of using appropriate controls cannot be overstressed! Controls validate immunochemical staining results and can quickly identify problems with reagents, test specimens, or an individual's immunochemical staining technique. The results of immunochemical staining are essentially meaningless unless the appropriate controls have been employed.

In any given immunochemical staining run, most laboratories routinely employ a positive tissue control, run alongside an identical section used as a reagent control.

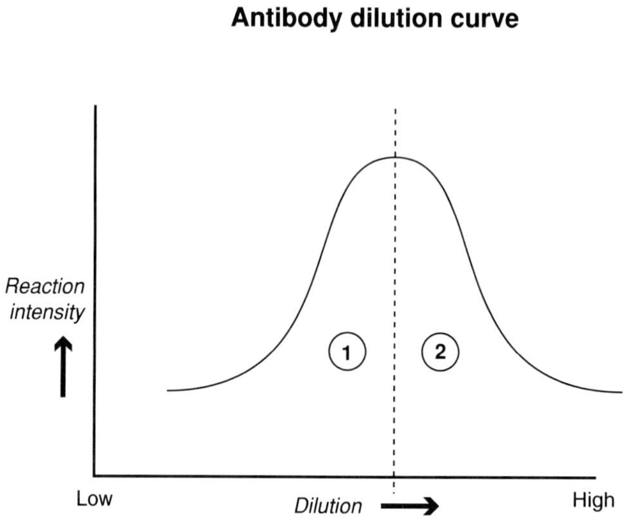

Figure 2. Determination of optimal dilution of antibody.
The poor reaction in area 1 is due to steric hindrance of the primary antibody accessing the antigen, due to the antibody concentration being too high. This is known as prozone, but is a rare phenomenon in immunochemical staining. The suboptimal reaction in area 2 is caused by insufficient primary antibody, i.e. the concentration of primary antibody is too low.

2.6.1 Positive and negative tissue controls

It must be stressed that control tissues must be subjected to *exactly* the same pre-treatments and immunochemical staining protocol steps as the test tissues in order to get an accurate result.

The inclusion of a positive control section known to contain the antigen in question is essential every time immunochemical staining is carried out. Without such a control, a negative result on the test material is meaningless because there is no guarantee that the reagents or kit are in good working order and have been applied in the correct order.

In contrast to the above, negative tissue controls do not contain the antigen in question. Any staining must therefore be from the primary antibody (either the wrong one has been used or from specific/nonspecific reactions) or nonspecific reactions intrinsic to the labeling system.

Although very rarely obtainable, knockout tissues make excellent negative controls. Mice, for example, can be genetically engineered to lack the protein (antigen) under investigation. A negative staining result from a knockout tissue combined with a positive result from a positive control tissue is a strong indication of specificity. However, both control types have to be from the same species for validity.

2.6.2 Reagent controls

The primary goal of reagent controls is to ensure that the primary and secondary antibodies recognize their target antigens. All other factors should be kept constant, for example the buffers used to dilute the antibodies, incubation time, and temperature.

Where possible, purchasing immunogen-affinity-purified primary antibodies and pre-absorbed secondary antibodies should greatly increase reagent specificity. Primary antibodies should first have their optimal dilutions determined, followed by testing on a panel of tissues known to be both positive and negative for the antigen in question.

An essential control when evaluating a new polyclonal antibody for demonstrating an antigen in an unknown location is to pre-absorb the primary antibody with an excess of its specific antigen, so that no antibody-binding sites are available to react with the tissue and staining does not occur (see *Fig. 3*). This may be regarded as the ultimate test for specificity. If staining does occur, then the staining must be due to a contaminating antibody or nonspecific interactions and not to the antigen–antibody interaction under investigation. For reasons of economy, absorption should be carried out at the highest possible dilution of the antibody compatible with consistent unequivocal staining, since the higher the concentration of antibody, the more antigen will be needed for neutralization.

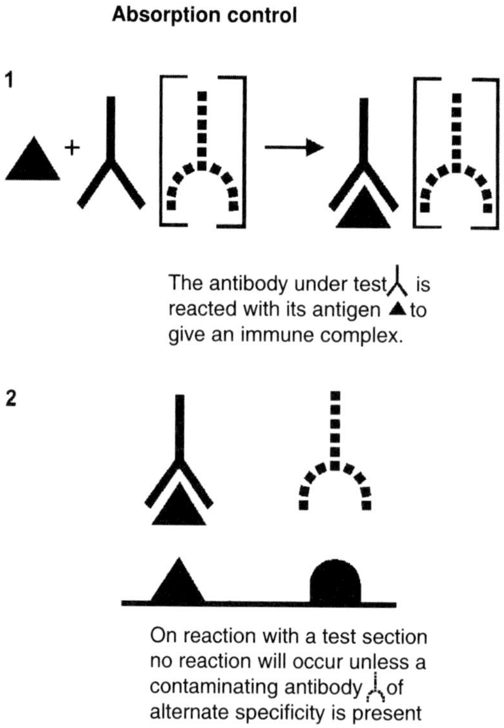
Absorption control
1
The antibody under test is
reacted with its antigen to
give an immune complex.
2
On reaction with a test section
no reaction will occur unless a
contaminating antibody of
alternate specificity is present

Figure 3. Example of absorption control for immunochemical staining.

However, if no antigen is available, or if financial or time constraints do not allow, then most laboratories replace the primary antibody under investigation with an isotype control or normal serum from the same species used to raise the primary antibody under investigation. An isotype control is defined as an antibody of the same isotype and raised in the same species as the primary antibody under investigation, but to an irrelevant protein, such as keyhole limpet hemocyanin. Any staining in the reagent control must be due to nonspecific reactions between the immunoglobulin and the tissue, or from reactions intrinsic to the labeling system. It is important when using an immunoglobulin fraction as a reagent control always to use it at the same immunoglobulin/overall protein concentration as that of the primary antibody under investigation. Similarly, both should be of the same age and obtained using the same purification methods. This is because soluble immunoglobulin aggregates present within the serum can exacerbate nonspecific staining, and the amount present is influenced by these factors.

If the primary antibody under investigation is monoclonal, then the only real option is to use an isotype control.

2.6.3 Internal controls

This type of control is a 'built-in' control and is the best possible control because the variables of tissue fixation, processing, and section treatment prior to staining are eliminated. Internal controls contain the target marker in the lesion to be identified and in the surrounding normal tissue elements. One example is the presence of S100 protein in melanoma and normal tissue elements such as peripheral nerves and melanocytes.

2.6.4 Control blocks

Control blocks should be selected from a case that expresses (is positive for) the antigen to be immunochemically stained. A negative control section should be cut from the same block.

Care should be taken when using stored or archived tissue sections as positive controls. Several publications have highlighted the dangers in the use of old tissue sections as positive controls. Some antigens in paraffin sections degrade during storage, probably due to oxidation. It is good practice to cut only enough sections from the control block to suffice for the week's work.

It is important that a section from the control block is stained with hematoxylin and eosin whenever a batch of control slides is cut to ensure that the section contains the lesion of interest.

2.7 Microscopic interpretation

Evaluation and interpretation of the finished product can be the most difficult aspect of immunochemical staining. Assessment of section quality regarding fixation, tissue processing, microtomy, and possible artifacts are all taken into consideration together with the demonstration and localization of the antigen.

Whoever interprets the immunochemical staining must know what the stain should look like in both normal tissue and tumor. Assessment of the pattern of staining is important. There are three patterns of specific staining that can be observed. These are cytoplasmic, nuclear, and surface. The amount of chromogenic precipitate on the labeling enzyme, and therefore the intensity of the reaction, is proportional to the amount of antigen present. Not all cells contain the same amount of antigen and they will therefore stain with varying intensity. Staining may also be focal or diffuse. Focal staining is localized in discrete areas of the cell, whereas diffuse staining may occupy larger areas of the cell and also adjacent cells.

Nonspecific background staining is usually associated with collagen and connective tissues, and although nonspecific background staining can be minimized by good immunochemical staining technique, it is difficult to abolish, particularly when using polyclonal antibodies. More problems in interpretation may arise due to endogenous enzyme activity in red blood cells and macrophages due to poor blocking techniques.

Interpretation is a comparison of the specific and nonspecific staining patterns of the test section with that of a control. Evaluation of a stained section should only be carried out when the tissue type, the basic histological procedures, the pre-treatment, and the immunochemical staining technique have all been taken into account.

Underfixation may preserve the antigen being investigated, but the morphology is likely to be poor and may cause difficulties in interpretation. Longer fixation will ultimately give better tissue morphology, but antigens may become altered, masked, depleted, or destroyed. Interpretation of staining patterns should be made based on adequately fixed areas of the section. Cells that have undergone autolysis, necrosis, or have been crushed may exhibit nonspecific staining. Only the staining pattern of viable cells should be considered for interpretation.

2.8 Background staining

The major causes of background staining in immunochemistry are hydrophobic and ionic interactions, and endogenous enzyme activity.

2.8.1 Hydrophobic interactions

Tissue hydrophobicity is increased by formaldehyde (and glutaraldehyde) fixation, leading to increased incidence of general background staining. All proteins are hydrophobic to some degree, with the side-chain amino acids phenylalanine, tryptophan, and tyrosine linking together in order to eradicate water. Aldehyde fixation increases hydrophobicity by linking reactive alpha and epsilon amino acids between adjacent proteins or within a single protein's tertiary structure. Connective tissues, squamous epithelium, and adipocytes (if not adequately removed during tissue processing) are especially prone. Such an increase in hydrophobicity is correlated to fixation temperature, duration, and pH, further enforcing the need for an optimized and standardized fixation regime.

Nonspecific background staining is most commonly produced because of the attachment of the primary antibody drawn nonimmunologically to highly charged groups present on connective tissue elements. Positive staining is due not to localization of the antigen but to nonspecific attachment of the primary antibody to connective tissues. Because the primary antibody is attached to connective tissue moieties, the subsequent labeling antibodies will be attracted not only to primary antibodies located on the specific antigen but also to the antibody bound to connective tissues. The most effective way of minimizing nonspecific staining is to add an innocuous protein solution, such as bovine serum albumin, to the section before applying the primary antibody. Other techniques are the addition of a detergent, e.g. Tween 20, to the washing buffer or using a diluent with a low ionic strength. The protein should saturate and neutralize the charged sites, thus enabling the primary antibody to bind only to its intended site.

Traditionally, nonimmune serum from the same animal species that produced the secondary antibody is used as a blocking serum, as this also has the effect of binding to any endogenous immunoglobulin in the tissue that shows reactivity with the secondary antibody. In practice, any animal serum or protein can be used for this purpose as long as the protein used as a block cannot be recognized by any of the subsequent antibodies used in the technique.

2.8.2 Ionic interactions

These occur when proteins of opposite net charges meet. Most IgG class antibodies have isoelectric points ranging from pH 5.8 to 7.3. Most will have a net negative surface charge when presented in the buffers normally used in immunochemical staining, i.e. in the pH range of 7.0–7.8. Interactions can therefore be expected if tissue proteins have a net positive surface charge. In general, ionic interactions can be reduced by the use of diluent buffers with higher ionic strength or by the addition of 0.1–0.5 M NaCl to the buffer normally used.

It should be noted that hydrophobic binding is reduced by the use of low ionic strength buffer. Ionic binding is reduced by the use of high ionic strength buffer. It can be seen that remedies to reduce one may aggravate the other!

2.8.3 Fc receptors

In frozen sections and cytological preparations, tissue receptors for the Fc portion of antibodies may give rise to additional problems. These Fc receptors present on several cell types such as macrophages and mono cytes as part of the natural immune defense mechanism are largely destroyed in paraffin-embedded tissues by formaldehyde fixation and tissue processing. If necessary, Fab or $F(ab)_2$ fragments of antibodies, which lack the Fc portion, should be used.

2.8.4 Endogenous peroxidase activity

Certain cells (in particular red blood cells) naturally contain this oxido-reductase enzyme. By applying hydrogen peroxide to the tissue prior to the staining procedure, endogenous peroxidases can be blocked and reversibly prevented from reacting with the chromogen applied as the last detection step (see Chapter 4, section 2.10.2, *Protocol 4*). This is due to the initial complex formed between the excess of peroxidase and hydrogen peroxide being catalytically inactive if no electron donor is present. Other blockers include 0.2% hydrochloric acid in ethanol, and methanol containing 1% sodium nitroferricyanide and 0.2% acetic acid. An alternative to blocking endogenous peroxidases is simply to use an alkaline phosphatase label instead of horseradish peroxidase.

Thick tissue sections (60 µm) in a free-floating format (i.e. not adhered to a slide; for example, gelatin-embedded) can be treated with a different method using sodium borohydate, which penetrates the thick tissue sections more efficiently than hydrogen peroxide.

Protocol 2

Blocking endogenous peroxidases in free-floating, thick tissue sections (60 µm)

Equipment and Reagents

- 1% (w/v) Sodium borohydrate solution
- Extraction cabinet
- Magnetic stirrer

Method

1. Wash the free-floating, thick tissue sections in PBS or TBS to remove excess fixative.

CARRY OUT STEPS 2–6 IN AN EXTRACTION CABINET (NO HUMIDITY)[a]

2. Prepare a fresh solution of 1% (w/v) sodium borohydrate in water: weigh 1 g of sodium borohydrate and add it to a suitable container containing 100 ml of double-distilled water in one movement. Close the lid immediately and stir gently using a magnetic stirrer.
3. Wait 5 min for the sodium borohydrate to dissolve in the water. Effervescence should be observed and the reaction will be exothermic.
4. Open the container and add the solution to the reaction vessels containing the sections. Incubate the sections with this solution for up to 10 min. Effervescence will be observed from the tissue due to oxygen release.
5. Wash the sections ten times in PBS or TBS until no effervescence is seen.
6. Continue with the immunochemical staining protocol.

Notes

[a]Sodium borohydrate is *extremely dangerous* as it is an explosive. It must be handled in a dry environment prior to being added to the water, as contact with water releases highly flammable gases; hence, the strict use of an extraction cabinet. Dispose of the used sodium borohydrate solution according to material safety data sheet (MSDS) guidelines.

Protocol 3

Blocking endogenous peroxidases in 4–10 μm thick tissue sections or cytological preparations

Please see Chapter 4, section 2.10.2, *Protocol 4*.

2.8.5 Endogenous phosphatase activity

This enzyme is found endogenously in liver, intestine, bone, placenta, leukocytes, and certain carcinomas. Applying levamisole at a concentration of 1–2 mM in the chromogen solution will block endogenous phosphatase in all tissues except the intestine, which exhibits a different isoform (8). Alternative endogenous alkaline phosphatase blocking solutions include Bouin's fixative and 20% acetic acid. An alternative to blocking intestinal phosphatase is simply to use a horseradish peroxidase label instead of alkaline phosphatase.

Protocol 4

Blocking endogenous phosphatases in 4–10 μm tissue sections or cytological preparations

Reagents

- Alkaline phosphatase chromogen solution containing 1 mM levamisole (Sigma, catalogue no. L9756)

Method

1. Use a standard immunochemical staining protocol, such as that given in Chapter 4, section 2.10.2, *Protocol 1*.
2. Apply the alkaline phosphatase chromogen solution containing 0.24 mg/ml (1 mM) levamisole.
3. Incubate for as long as necessary to see specific staining.

Protocol 5

Blocking endogenous intestinal phosphatases in 4–10 μm tissue sections or cytological preparations

Reagents

- Absolute methanol containing 20% acetic acid (v/v)

Method

1. Use a standard immunochemical staining protocol, such as that given in Chapter 4, section 2.10.2, *Protocol 4*.
2. Apply absolute methanol containing 20% acetic acid (v/v) for 5 min instead of the H_2O_2 (step 11).
3. Continue with the immunochemical staining protocol.

2.8.6 Biotin

Liver, kidney, and lymphoid tissues contain high levels of this vitamin B7. It can also be found in adipose tissue and central nervous tissues in low amounts.

Biotin is often used in immunochemical staining methods to amplify the signal (see Chapter 4, section 2.10.2, *Protocol 4*). It is therefore important to block endogenous biotin when an ABC amplification system is being used, since the avidin, when applied to the tissue or cells during the amplification step, can potentially bind to the endogenous biotin. To block endogenous biotin, it is recommended pre-incubating the tissue sections in 0.01% (w/v) avidin (to bind to the endogenous biotin), followed by 0.001% (w/v) biotin (to block any vacant biotin-binding sites on the avidin).

Protocol 6

Blocking endogenous biotin in 4–10 μm tissue sections

The following steps should be performed before incubation with the biotinylated antibody.

Reagents

- PBS or TBS containing 0.01% (w/v) avidin (Sigma, catalogue no. A9275)
- PBS or TBS containing 0.001% (w/v) biotin (Sigma, catalogue no. B4501)
- PBS or TBS

Method

1. Prepare fresh solutions of PBS or TBS containing 0.01% (w/v) avidin, and PBS or TBS containing 0.001% (w/v) biotin.
2. Apply the PBS or TBS containing 0.01% (w/v) avidin to the tissue sections for 10 min.
3. Rinse the slides in PBS or TBS.
4. Apply PBS or TBS containing 0.001% (w/v) biotin to the tissue sections for 10 min.
5. Rinse the slides in PBS or TBS.
6. Continue with the immunochemical staining protocol.

A very effective biotin blocking solution for heavily biotinylated sites – which can be very difficult to block using the standard blocking solutions – can also be made in house (9).

Protocol 7

Blocking heavily biotinylated sites in 4–10 μm tissue sections

The following steps should be performed before incubation with the biotinylated antibody.

Reagents

- 200 ml Distilled water containing two egg whites (solution A)
- Distilled water containing 0.2% (w/v) biotin (Sigma, catalogue no. B4501) (solution B)
- PBS or TBS

Method

1. Apply solution A to the tissue sections for 20 min.
2. Rinse the slides in TBS.
3. Apply solution B to the tissue sections for 20 min.
4. Rinse the slides in TBS.
5. Continue with the immunochemical staining protocol.

2.8.7 Autofluorescence

Autofluorescence is the ability of a tissue or cellular component to fluoresce naturally. This occurs independent of the binding of immunochemical staining reagents conjugated to fluorochromes. Autofluorescence can lead potentially to false-positive staining. Three types of autofluorescence are common:

- Elastin and collagen autofluorescence
- Lipofuscin autofluorescence
- Aldehyde (fixative)-induced autofluorescence

Elastin and collagen autofluorescence

Elastin and collagen are common components of blood vessels. Amongst other fluorophores, elastin contains a naturally occurring cross-linking tricarboxylic amino acid with a pyridinium ring (10). Collagen possesses a similar fluorophore.

Cowen *et al.* (11) used a dye called pontamine sky blue to successfully quench autofluorescence in carotid arteries and mesenteric vessels using a fluorescence resonance energy transfer method. This method only works when using a fluorescein isothiocyanate (FITC) fluorochrome, as the pontamine sky blue shifts the fluorescent emission of elastin and collagen from green to red.

Protocol 7

Quenching autofluorescence due to elastin and collagen in tissue sections (11)

Reagents

- PBS or TBS containing 0.5% (w/v) pontamine sky blue

Method

1. Use a standard immunochemical staining protocol, such as that given in Chapter 4, section 2.10.2, *Protocol 4*.
2. Apply PBS or TBS containing 0.5% (w/v) pontamine sky blue for 10 min at the step before incubation with the reagent conjugated to the FITC fluorochrome.
3. Rinse the slides in PBS or TBS for 5 min.
4. Continue with the immunochemical staining protocol.

Lipofuscin autofluorescence

Naturally fluorescent lipofuscin pigment accumulates with age in the cytoplasm of numerous cell types, including those of the central nervous system (12, 13). Commonly referred to as ageing pigment, it is yellow-brown in appearance and accumulates within lysosomes due to peroxidation of lipids (6). It has an excitation and emission spectrum similar to that of FITC.

Protocol 8

Quenching autofluorescence due to lipofuscin in tissue sections (12, 13)

Reagents

- 70% (v/v) Ethanol containing 1% (w/v) sudan black B (stirred in the dark for 2 h)

Method

1. Use a standard immunochemical staining protocol, such as that given in Chapter 4, section 2.10.2, *Protocol 4*.
2. Apply 70% (v/v) ethanol containing 1% (w/v) sudan black B for 10 min at the step before incubation with the reagent conjugated to the fluorochrome.
3. Quickly rinse the slides ten times in TBS or PBS.
4. Continue with the immunochemical staining protocol.

Aldehyde induced autofluorescence

Autofluorescent compounds are formed between aldehyde fixatives and proteins or amines, with glutaraldehyde fixation giving a greater degree of autofluorescence than formaldehyde due to extensive cross-linking. Aldehyde-induced autofluorescence is more diffuse and more generalized than specific staining, its intensity increasing with incubation temperature and duration (see Chapter 4, section 2.1, for further information on fixation).

An effective means of avoiding aldehyde-induced autofluorescence is simply to avoid using an aldehyde fixative, although this can have certain disadvantages. See Chapter 4, section 2.1, for a discussion of alternative fixatives.

Protocol 9

Quenching autofluorescence in aldehyde-fixed tissue sections or cytological preparations (14)

Equipment and Reagents

- 0.1% (w/v) Sodium borohydrate solution
- Extraction cabinet
- Magnetic stirrer

Method

1. Use a standard immunochemical staining protocol, such as that given in Chapter 4, section 2.10.2, *Protocol 4*.

CARRY OUT STEPS 2–5 IN AN EXTRACTION CABINET (NO HUMIDITY).[a]

2. Prepare a fresh solution of 1% (w/v) sodium borohydrate in water: weigh 100 mg of sodium borohydrate and add it to a suitable container containing 100 ml of double-distilled water in one movement. Close the lid immediately and gently stir using a magnetic stirrer.
3. Wait 5 min for the sodium borohydrate to dissolve in the water. Effervescence should be observed and the reaction will be exothermic.

4. Open the container and apply this solution to the tissue section or cells (while still effervescing)[b]. For cell monolayers, incubate twice for 4 min each using fresh solution each time. For paraffin-embedded sections, incubate three times for 10 min each using fresh solution each time. Effervescence will be observed from the tissue due to oxygen release.
5. Rinse three times for 5 min each in PBS or TBS to remove all traces of sodium borohydrate.
6. Continue with the immunochemical staining protocol

Notes

[a]Sodium borohydrate is *extremely dangerous* as it is an explosive. It must be handled in a dry environment prior to being added to the PBS or TBS, as contact with water releases highly flammable gases; hence, the strict use of an extraction cabinet. Dispose of the used sodium borohydrate solution according to MSDS guidelines.

[b]Sodium borohydrate reduces aldehyde-induced autofluorescence by reducing Schiff bases ($R_1HC{=}NR_2$) that are formed from the reaction between the aldehyde and NH_2 groups.

2.8.8 Formaldehyde (formalin) pigment

Specimens fixed in acidic formaldehyde can demonstrate formaldehyde pigment, a brown-black deposit created when acidic formaldehyde reacts with blood. This pigment is related to the acid hematins and can be mistaken for immunochemical staining when diaminobenzidine (DAB) is used as the chromogen (substrate) (see Chapter 4, section 2.1.1). Formation of formaldehyde pigment can be almost eliminated by fixing in neutral buffered formaldehyde (pH 7.0) (15).

3. TROUBLESHOOTING

Troubleshooting immunochemical staining procedures is not an easy task. With so many variables, it is often difficult to know exactly where to begin. The best approach is to start with the simple and progress to the more technically demanding issues later:

- Start by ensuring that none of the steps of the immunochemical staining protocol have been omitted, according to whatever detection system is being employed.
- Ensure that all relevant blocking steps have been performed.
- Consult the datasheet for the primary antibody to verify that it will indeed work in the specimen type that is being used; for instance, frozen sections, paraffin-embedded sections, or cytological preparations. The datasheet for the primary antibody may also give extremely helpful information regarding optimal specimen fixation conditions and antigen retrieval.
- Check that all of the reagents are compatible with each other; for instance, ensure that the secondary antibody detects the immunoglobulin species and subclass of the primary antibody.

- If no apparent problems can be found with the above, check that all of the reagents used have been made up correctly; for instance, that the buffer constituents and pH are correct.
- Check that none of the reagents has gone out of date or has been stored incorrectly.

Table 1 is designed to give the user insight into the possible causes of error during an immunochemical staining procedure and the possible solutions, grouped by the observed staining pattern seen in the test specimen and the positive and negative tissue/reagent controls.

Table 1. Troubleshooting immunochemical staining procedures

Possible cause	Possible remedy
Weak or absent immunochemical staining in both control and test specimens, with minimal or very little background staining	
Paraffin section (cytological preparation if PEG protected) is not dewaxed properly	Ensure an adequate time in the dewaxing solution
Tissues exposed to too high a temperature during embedding or paraffin section drying	The wax bath and oven temperature should not exceed 60°C (see Chapter 4, section 2.2.1)
Fixative used is incompatible with the antigen	Refer to the antibody datasheet for possible fixative recommendations
Reagent(s) omitted or applied in wrong order	Follow the immunochemical staining protocol carefully
Incompatible reagents	Ensure that the primary antibodies detect antigen in the species and sample type (frozen, formaldehyde-fixed, paraffin embedded, etc.) being studied
	Ensure that the secondary antibodies detect the primary antibody subclass and isotype
	If using an ABC system, ensure that the secondary antibody is biotinylated
	Ensure that the chromogen is compatible with the immunoenzyme label
Reagent(s) made up incorrectly	Follow the manufacturer's instructions and protocol recipes carefully
Reagent(s) too old or degraded	Observe the manufacturer's expiration dates and storage conditions

Reagent(s) too old or degraded *(contd)*	Ensure that frozen antibodies are not subjected to excessive freeze–thaw cycles (see Chapter 2, section 2.2.9)
	Ensure that conjugated antibodies are not frozen (see Chapter 3, section 2.6)
Reagent(s) used at too low/high a concentration	Perform dilution ranges for the antibodies used at each stage to find the optimum (see section 2.5.2). Note: at too high a concentration, antibodies may exhibit 'prozone' effects (see section 2.5.2)
	Observe the manufacturer's recommended working dilutions/concentrations for commercial reagents
Reagent(s) not incubated for long enough	Increase the incubation times of reagents, in particular for the primary antibody
Reagent(s) not incubated at a high enough temperature	Increase the incubation temperature by a few degrees
	Allow all reagents to warm to room temperature before starting the incubation
Antigen not present at detectable levels	Use a more sensitive detection system (see Chapter 4, *Table 4*)
	Increase the incubation times of reagents, in particular for the primary antibody
	Antigen may be masked by formaldehyde fixation, so perform antigen retrieval (see section 2.4, this chapter, and Chapter 4, section 2.5) if not already done
	If antigen retrieval has already been performed, check that the reagents have been made up correctly, increase retrieval times, increase heat during HIER, check for defective or underactive performance of the enzyme during enzymatic retrieval, or try another retrieval technique
	Use a different (often noncross-linking) fixative
Wrong buffer/buffer constituents/pH used.	Avoid PBS when using phosphatase labels
	Avoid sodium azide when using peroxidase labels. Note: diluent pH can affect antibody affinity for the antigen
	Detergent in buffers can remove some membrane-bound antigens

Dissociation of reagents during buffer washes	Use antibodies at lower dilutions
	Replace primary antibody with antibody of the same specification but with a higher affinity for the antigen
Enzyme or chromogen precipitate is soluble in nuclear counterstain, dehydrating solution, or mounting media	Use a nonalcohol-containing counterstain such as Mayer's
	Use a nonalcohol-soluble chromogen such as DAB
	Use aqueous mounting medium
Photobleaching occurs when using fluorescent labels	Carry out all immunochemical staining incubations in the dark (see Chapter 4, section 2.10.2)
Using a mountant containing glycerol when using phycobiliprotein fluorescent labels	Use a nonglycerol-containing mountant (see Chapter 4, section 2.7)
Nuclear counterstain is too heavy in the case of nuclear antigens	Counterstain for a shorter length of time and/or differentiate for longer in the case of regressive hematoxylin (see Chapter 4, section 2.6)
Positive control specimen demonstrates specific staining with minimal or very little background staining. Weak or absent immunochemical staining in test specimen, with varying degrees of background staining	
Antigen not present or present at too low a concentration in the test specimen	Use a more sensitive immunochemical detection technique (see Chapter 4, Table 4)
Test specimen is not adequately fixed	Fix for an extended length of time
	Fix smaller pieces of tissue to aid fixative penetration
Test specimen fixed for excessive length of time in cross-linking fixative	Optimize and standardize fixation times
	Perform antigen retrieval
Test specimen exhibits necrotic/damaged areas or crush artifacts	Cut sections with a sharp microtome or cryostat blade
	Ignore such areas
General background staining demonstrated in both test and control specimens	
Excessive incubation times of reagents, especially primary antibody and chromogen	Optimize and standardize incubation times
	Refer to product literature for suggested incubation times

Excessive reagent concentrations	Optimize working dilutions or concentrations for each reagent
	Observe manufacturer's recommended working dilutions or concentrations for commercial reagents
Inadequate rinsing of slides	Rinse slides adequately
Secondary antibody binding to endogenous immunoglobulins and/or antigens in specimen	Use a primary antibody raised in a species different to that of the test specimen
	Use a secondary antibody pre-adsorbed against normal serum or tissue extract from the species of the test specimen
Incorrect blocking serum used	Use serum from the species in which the secondary antibody was raised
Inadequate or absent blocking of endogenous peroxidases and/or phosphatases	Perform appropriate block(s) (see Chapter 4, section 2.10.2, *Protocol 4*, note j; and sections 2.8.4 and 2.8.5, this chapter)
Endogenous biotin when using an ABC immunochemical staining method (common in kidney and liver specimens)	Perform a biotin block using an excess of avidin followed by an excess of biotin. Use a proprietary blocking kit or use the recipe given in *Protocol 6*
	Use a nonbiotin immunochemical staining method
Drying of the specimen before fixation or during immunochemical staining	Immerse specimen into fixative immediately after being dissected
	Carry out all immunochemical staining incubations in a humidified environment (see Chapter 4, section 2.10.2)

General background staining demonstrated in negative reagent control and test specimen. Positive and negative control specimens demonstrate expected staining with minimal or very little background staining

Test specimen not adequately fixed	Fix for an extended length of time
	Fix smaller pieces of tissue to aid fixative penetration
Test specimen is fixed for excessive length of time in cross-linking fixative, increasing tissue hydrophobicity	Optimize and standardize fixation times
	Perform antigen retrieval
Test specimen exhibits necrotic or damaged areas or crush artifacts	Cut sections with a sharp microtome or cryostat blade
	Ignore such areas

Excessive or patchy application of tissue adhesive to slides, therefore adhering immunochemical reagents	Observe manufacturer's recommended coating procedure when using commercial solutions
Tissue sections cut too thickly, 'trapping' immunochemical reagents	Aim for 4 μm thick sections and certainly no thicker than 10 μm (see section 2.2)
General background staining demonstrated in negative reagent control specimen. The positive control, negative control, and test specimens demonstrate the expected staining with minimal or very little background staining	
Negative-control serum is too concentrated	Dilute the negative-control serum to the immunoglobulin concentration of the primary antibody (or whole protein concentration if the primary antibody is unpurified)
Antibodies present in the negative-control serum are cross-reacting with tissue components in the test specimen, or bacterial/fungal infection of the negative-control serum	Use a new batch of negative-control serum
Leukocyte staining is demonstrated in all or some specimens (usually frozen sections)	
Binding of the Fc region of the primary antibody to Fc receptors on leucocytes	Use Fab or $F(ab)_2$ primary antibody fragments
	Add detergent to buffers to dissolve Fc receptors
Specific but unexpected staining is demonstrated in all or some specimens	
Contaminating antibodies are present in the polyclonal primary antibody solution that detect a different target to the desired one	Use an immunogen-affinity-purified version of the antibody (or at least a protein A or G (or protein Y for chicken antibodies)-affinity-purified version)
Positive staining on negative-control section	
Primary antibody has not been omitted from section used as the negative control	Ensure that the primary antibody has been omitted
Contamination of the secondary antibody during a previous procedure. Labeling antibody now binds to a tissue epitope	Ensure that a fresh pipette tip is used every time, and never reuse reagents
Endogenous biotin. This can be a problem when using avidin–biotin detection systems on tissues such as kidney and breast	Perform a biotin block using an excess of avidin followed by an excess of biotin. Use a proprietary blocking kit or use the recipe given in *Protocol 6*
Poor tissue morphology	
Loss of cell and tissue integrity, usually associated with poor fixation and paraffin processing	Reprocess the paraffin block or select another block on which to carry out the tests
Excessive proteolytic enzyme digestion or over-retrieval (HIER)	Check the concentration of enzyme and duration of digestion. Modify the HIER technique

If no solution can be found after consulting this table, it is often worthwhile contacting the technical department of the primary antibody supplier to see whether they can be of assistance. Failing this, the user may have the luxury of being able to hand over their primary antibody to a fellow colleague to see what results they can obtain. This is especially useful when they carry out the work in a different laboratory, where the laboratory's own protocols, reagents, and controls are used. They should immunochemically stain the user's own test specimen and controls alongside their own, subjecting them to the same immunochemical staining protocol and pre-treatment. If they manage to obtain the desired staining pattern, then comparing and contrasting the techniques, reagents, test specimens, and controls used should provide (or move the user very close to) the answer.

In view of the often complex nature of immunochemical techniques and as a further validation of laboratory procedures, participation in an external quality assurance scheme is recommended. The UK National External Quality Assessment Scheme for Immunochemistry (UK NEQAS-ICC) assessments takes place at 3-monthly intervals throughout the fiscal year. Currently, laboratories are able to participate in up to six different modules, depending on their service commitments and specialized areas of interest. These modules are:

1. The general pathology module, catering for routine markers used by the majority of pathology departments offering a routine immunochemistry service.
2. The breast hormonal receptor module, catering for laboratories routinely demonstrating estrogen and progesterone receptors on paraffin-embedded tissues.
3. The breast HER2 module, catering for laboratories routinely testing for HER2 on paraffin-processed tissues.
4. The lymphoma module, catering for laboratories with a special interest in lymphoma pathology.
5. The neuropathology module, catering for the markers common to most neuropathology laboratories.
6. The cytology module, catering for markers commonly requested on cytological preparations.
7. The alimentary tract pathology module, catering for laboratories specializing in this area of pathology.

More information can be obtained from the Scheme Manager at UK NEQAS-ICC, Suite 3/22 Hamilton House, Mabledon Place, London WC1H 9BB, UK.

4. REFERENCES

★★ 1. **Helander KG** (1994) *Biotech. Histochem.* 69, 177–179. – *Explains the methods of formalin fixation and the problems it causes with immunocytochemical investigations.*

★★ 2. **Miller RT** (2001) *Technical Immunohistochemistry: Achieving Reliability and Reproducibility of Immunostains.* Society for Applied Immunohistochemistry Annual Meeting, 8 September, 2001, NY, USA. – *Describes the effects of various commonly used decalcification agents on tissue antigens.*

3. **Athanasou NA, Quinn J, Heryet A, Woods CG & McGee JO** (1987) *J. Clin. Pathol.* **40**, 874–878.

4. **Matthews JB & Mason GI** (1984) *Histochem. J.* **16**, 771–787.

5. **Mukai K, Yoshimura S & Anzai M** (1986) *Am. J. Surg. Pathol.* **10**, 413–419.

★★ 6. **Huang SN, Minassian H & More JD** (1976) *Lab. Invest.* **35**, 383–390. – *Describes the effect of proteolytic enzyme digestion techniques on paraffin embedded material.*

7. **Brozman M** (1980) *Acta Histochem.* **67**, 80–85.

8. **Boenisch T (ed.)** (2001) *Immunohistochemical Staining Methods Handbook*, 3rd edn. DakoCytomation, CA, USA.

9. **Miller RT** (1997) *Appl. Immunochem.* **5**, 63–66.

10. **Deyl Z, Macek K, Adam M & Vancikova O** (1980) *Biochim. Biophys. Acta*, **625**, 248–254.

11. **Cowen T, Haven AT & Burnstock G** (1985) *Histochemistry*, **82**, 205–208.

12. **Schnell SA, Staines WA & Wessendorf MW** (1999) *J. Histochem. Cytochem.* 47, 719–730.

13. **Yin D** (1996) *Free Radic. Biol. Med.* **21**, 871–888.

14. **Clancy B & Cauller LJ** (1998) *J. Neurosc. Methods*, **83**, 97–102.

15. **Bancroft J & Stevens A (eds.)** (1996) *Theory and Practice of Histological Techniques*, 4th edn. Churchill Livingstone, NY, USA.

CHAPTER 10

Automated immunochemistry

Emanuel Schenck

1. INTRODUCTION

Automation of immunochemical techniques in the laboratory has taken on considerable momentum since it first gained recognition in 1992 (1). Continuous progress in this area is fuelled by several factors, including improved consistency and quality of automated results, a growing scarcity of highly trained technicians required for manual staining, and the need for faster turnaround times in delivering results.

We have witnessed considerable technological progress in the last decade and it is therefore appropriate to expect that automated technology will emerge as an enabling tool for various types of laboratories in the future. Modern automated immunochemical platforms are capable of taking over every task in the immunohistochemical laboratory, performing essential and tedious steps such as antigen retrieval or dewaxing of paraffin slides. Most scientists would agree that having this technology available in the immunochemical staining laboratory is a significant advantage.

Although the benefits of automation have not been systematically evaluated and documented in a large number of scientific publications (2–4), the recent spread of automated systems into all areas of immunochemical practice indicates that manual methods have been displaced as the 'gold standard' for various immunochemical applications and assays. In the absence of detailed performance measures for the various systems, it is essential to evaluate the individual platforms in terms of overall quality of staining produced for various assays.

As there are quite a number of different automated stainers on the market, selection of the appropriate platform is not an easy task. Although similarity of automated protocols to manual methods may be important for some laboratories, it is probably wise not to regard this as the main

Immunohistochemistry: *Methods Express* (S. Renshaw, ed.)

criterion for selection of a platform. It can be expected that the technology will continue to provide solutions that go beyond the automated adaptation of manual methods. This chapter is intended to review the merits of the latest technological advances that enable faster throughput of immunochemical staining methods and to provide some guidance on the drawbacks associated with some of the current technologies and commercially available systems.

It must be noted that this chapter does not contain any protocols. This is due to the fact that automated immunochemical staining machines follow protocols that are identical to those employed in manual immunochemical staining methods, with the various steps being programmed in individually or as a pre-set package by the operator before staining commences. Such protocols can be found in Chapter 4, section 2.10.2.

1.1 Defining the needs

The task of choosing the optimal platform for a particular laboratory is simplified considerably if the needs are clearly defined in advance. Some of the relevant questions to review before selection are listed below:

- Is it essential for the operator to have a system that has the highest throughput (as measured in slides/hour) performing immunochemical staining in large batches using relatively few different protocols, or should the laboratory be equipped with a system that allows the highest degree of experimental flexibility to cope with numerous different protocols and modifications?
- Are additional capabilities of the system such as *in situ* hybridization important?
- Is the aim to provide walk-away operation of immunochemical staining with the highest degree of automation and standardization in order to free up valuable time for the technical staff to perform other tasks, or is the amount of 'hands-on' time (for protocol selection and modification, slide loading, etc.) required of the operator less important?
- Is the highest degree of standardization of the procedures desirable?
- Is the ability of the vendor to provide optimized reagents for all desired applications essential, or is overall reagent cost a major factor for the day-to-day operation of the laboratory?

For many laboratories it can be expected that any one platform will not fulfill all of the requirements. This often means that operators choose to purchase several platforms to cover the entire spectrum of tasks.

2. METHODS AND APPROACHES

Automated systems are generally employed to increase throughput, decrease slide-to-slide variability, and introduce a new level of reproducibility and standardization. Other desired endpoints include improved traceability and self-checking of critical steps.

2.1 Overview of automated platforms for immunochemical staining

Various types of automated slide-staining systems are currently on the market. Most systems that are currently installed in histology laboratories are used to perform only immunochemical staining. Other systems are used to perform both immunochemical staining and *in situ* hybridization. The latter technique demands that the system provides both a means to heat slides to perform denaturation and hybridization steps and the ability to prevent sample degradation by desiccation. A number of automated immunochemical staining platforms can also be used to perform tinctorial stains. In the past, slide-to-slide variability has been an issue for tinctorial stains performed on automated immunochemical staining systems. Many vendors have now adapted their staining protocols to overcome these problems of consistency.

2.2 System contrasts

There are considerable differences between the various automated systems with regard to:

- Overall configuration of the system.
- Reagent application method.
- The ability of the instrument to apply heat to individual slides within a narrow temperature range

These and other parameters have to be evaluated before individual platforms are looked at in more detail. A summary of some of the basic features of selected immunochemical stainers is presented in *Table 1*.

2.2.1 Overall configuration of the system

Automated immunochemical stainers fall into two types of category with regard to their basic configuration: array stainers and rotary stainers. In array stainers, the slides are arranged in a matrix configuration of rows and columns (see *Fig. 1*).

Table 1. Commercially available platforms: basic features

Vendor*	Product name	Configuration	Operating modes	Reagent application method	Advertised staining capabilities**
BioCare Medical	Nemesis 7200	Array	Barcode & open	Rinsed probe	IC
BioGenex Laboratories	*i*6000	Array	Barcode & open	Pipette tips	IC, ISH
DakoCytomation	Autostainer Plus	Array	Barcode & open	Rinsed probe	IC
DakoCytomation	Eridan	Array	Barcode & open	Rinsed probe	IC
Lab Vision	Autostainer 480	Array	Open	Rinsed probe	IC
Lab Vision	Autostainer 720	Array	Open	Rinsed probe	IC
Tecan	Freedom EVO	Array	Open	Pipette tips	IC, ISH
Ventana Medical Systems	BenchMark XT	Rotary	Barcode	Disposable dispenser	IC, ISH
Ventana Medical Systems	BenchMark LT	Rotary	Barcode	Disposable dispenser	IC, ISH
Ventana Medical Systems	NexES IHC	Rotary	Barcode	Disposable dispenser	IC
Vision BioSystems	Bond Max	Array	Barcode & open	Rinsed probe	IC, ISH
Vision BioSystems	Bond X	Array	Barcode & open	Rinsed probe	IC
Zymed Laboratories	Mozaic	Array	Open	Rinsed probe	IC, ISH

*Further information about products can be obtained from the vendor websites: BioCare Medical, www.biocare.net; BioGenex Laboratories, www.biogenex.com; DakoCytomation, www.dako.dk; Lab Vision, www.labvision.com; Tecan, www.tecan.com; Ventana Medical Systems, www.ventanamed.com; Vision BioSystems, www.vision-bio.com; Zymed Laboratories, www.invitrogen.com.
**IC, Immunochemistry; ISH, *in situ* hybridization.

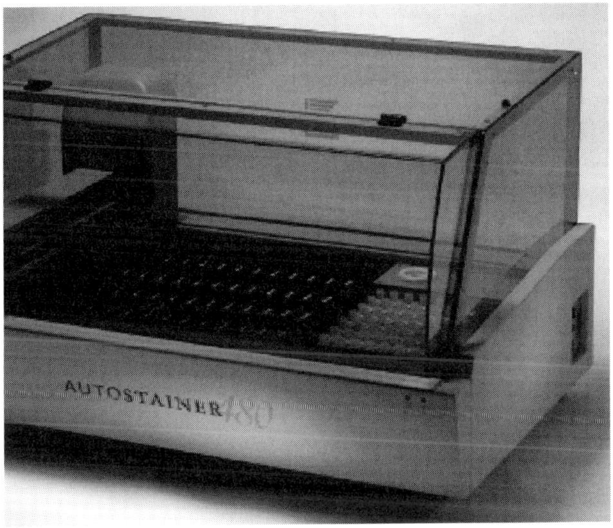

Figure 1. The Lab Vision Autostainer 480 array stainer.

One advantage of these stainers is that slides can be removed as staining procedures are completed. Some systems also offer continuous batch processing, allowing batches of slides to be stopped and started independently. Array stainers generally allow the operator to mimic manual procedure and offer the highest degree of experimental flexibility with regard to protocol design and the use of reagents from different vendors.

Rotary stainers, on the other hand, use circular arrangements of reagents and slides. A carousel is used to rotate reagents into position before they are applied to slides situated on a circular, stationary platform (see *Fig. 2*).

Although rotary stainers limit the operator's ability to modify staining protocols to some extend, they do allow simultaneous execution of very different protocols. In rotary stainers, access to slides during the staining procedure is generally not as easy as in array stainers. This could be perceived as a disadvantage in laboratories that require staggered workflows.

Staining platforms are designed to be either closed or open systems. Closed systems restrict the operator's ability to use reagents from sources other than the platform's distributor. Depending on the overall use of the system within the laboratory, this can have important implications on overall running costs. The cost per slide is typically three to four times that of a manual run. Rotary stainers are usually closed systems and use bar-coded slide labels to determine the appropriate staining protocol. Open

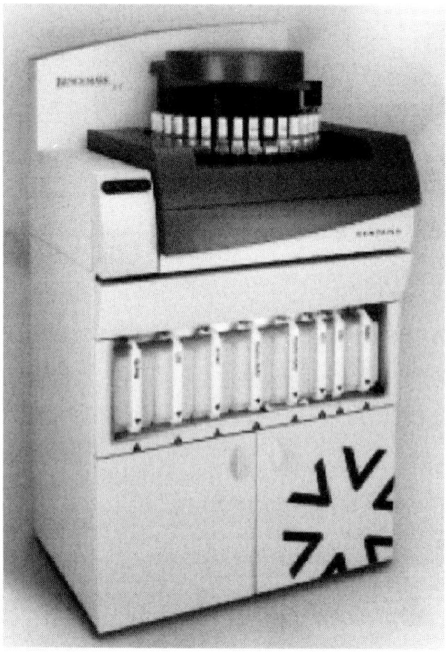

Figure 2. The Ventana Benchmark XT rotary stainer.

systems are more or less designed to allow use of reagents obtained from sources other than the instrument's distributor. This will often result in greater experimental flexibility in designing the individual staining assays. On the other hand, selection of an open system may have implications for the overall hands-on time necessary to operate the platform, and standardization of the procedures may not always be optimal. Array stainers are usually open systems, requiring the operator to select protocols and place slides on the array platform and then instruct the operator where to place the reagents. A number of laboratories that perform more routine immunochemical staining work prefer closed systems for their assays. The demand for open systems tends to come from research laboratories. It is advisable to evaluate individual platforms and to determine in detail whether they offer the desired experimental flexibility.

2.2.2 Reagent application method

Another aspect to consider with automated technologies is the means by which liquids are applied to and displaced from the slides. Different reagent-delivery techniques can lead to different problems.

Some systems exploit capillarity forces to ensure the even spread of reagent solutions over the slide. In these systems, the slides are placed in a vertical position and a gap between the slide surface and a slide cover is used to draw reagents over the surface. Other systems employ flat immunochemical staining: slides are positioned horizontally in the incubator and reagents are delivered by syringe-like or probe-type delivery systems or through disposable pipette tips. Problems with capillarity include the improper filling and draining of the capillary gap and are often related to variations in gap sizes for different types of sections. Difficulties such as these can be avoided if the sections are mounted close to the bottom of the slide.

Rotary stainers dispense reagents through syringe-like or printer cartridge-like dispensers. The reagent container is rotated into position above the slide and the syringe-like system is used to dispense a predetermined volume of the reagent on to the slide.

Probe-type dispensers, which are similar to devices in chemical analyzers, are also used in a number of different platforms. Reagent carry-over can, however, pose a problem with these if the rinsing mechanisms are not adequate. Some probe-type systems such as the Autostainer from DakoCytomation allow precise definition of reagent volume and drop zone. The drop zone for these stainers can be the top, middle, or bottom of the slide or a combination of zones. The reagent volume can be adjusted precisely for each of the zones.

Yet another method of dispensing reagent on to a slide is to make use of disposable pipette tips, which have to be loaded and ejected by a robotic mechanism. Loss of tips during the staining procedures and other problems related to pipette loading and ejection can occur in these stainers.

2.2.3 Ability of the instrument to apply heat to individual slides within a narrow temperature range

Individual and other heating methods are features of some platforms, which enable heat-mediated 'antigen retrieval' and deparaffinization. As antigen retrieval is often the most critical step in the success of an immunochemical staining procedure, these systems provide a significant advantage over manual methods in terms of standardization. Slide heating is also a prerequisite for dual functionality of platforms as truly automated stainers for immunohistochemistry and *in situ* hybridization. A further advantage of this feature is that, by elevating the incubation temperature of primary antibodies, equilibrium can be achieved without the need to use extended incubation at 4°C. Some platforms have unique features such as, for example, the units from Ventana Medical Systems: these platforms use a solution known as Liquid Coverslip, which provides a uniform distribution of reagents and evaporation control on the slide.

2.3 Other special features

Other technologically significant aspects of automated platforms revolve around the software and peripherals supplied with the platforms. The software of modern immunohistochemical platforms is generally user-friendly. However, configuration of the system, selection of the protocol, loading, etc. in some of the less-advanced systems can cost considerable time. Many platforms can produce audit trails, which are increasingly important in the controlled environment of certain types of laboratories. The workflow in many histology laboratories tends to be more and more defined by automated labeling systems and many platforms follow this trend using bar-coded labels, which determine the staining procedures. The ability to run multiple staining units with only one computer may also be considered an advantage of some of the platforms.

Selected special features of selected immunohistochemical stainers are presented in *Table 2*.

2.4 System running costs

Another major consideration is the overall running cost (cost per slide) of the system. The difference in cost per slide between manual technique and automated systems varies significantly for the different platforms. For example, cost per slide is considerably higher for so-called 'closed' systems requiring reagents provided by the distributor of the platform. Use of other so-called 'open' systems will often not result in a notable increase in cost per slide when compared with that of manual technique.

2.5 System failure safeguards

When you take the overall number of slides and cost of reagents for each individual run into consideration, it becomes important to eliminate any factors that might interfere with the ability of the system to complete the run. Many systems do not complete a run if the power is interrupted, making uninterrupted power supplies essential to avoid this happening. Features designed to avoid human error can be considered a major advantage in some of the platforms. For example, special features of these platforms include safeguard mechanisms against reagent depletion by calculating the total amount of reagent required before a run is initiated. These systems will refuse to initiate a run if the reagent volume is not sufficient to complete all of the staining procedures and it is therefore common practice for technicians to overfill reagents slightly. Bar-coding

Table 2. Commercially available platforms: special features

Vendor	Product name	Slide capacity	Slide identification*	Reagent container identification*	Slide heating
BioCare Medical	Nemesis 7200	84	Yes	Yes	No
BioGenex Laboratories	*i*6000	60	Yes	Yes	No
DakoCytomation	Autostainer Plus	48	Yes	No	No
DakoCytomation	Erldan	64	Yes	Yes	Yes
Lab Vision	Autostainer 480	48	Yes	Yes	No
Lab Vision	Autostainer 720	72 (84)	Yes	Yes	No
Tecan	Freedom EVO	48	No	No	Yes
Ventana Medical Systems	BenchMark XT	30	Yes	Yes	Yes
Ventana Medical Systems	BenchMark LT	20	Yes	Yes	Yes
Ventana Medical Systems	NexES IHC	20	Yes	Yes	Yes
Vision BioSystems	Bond-Max	30	Yes	Yes	Yes
Vision BioSystems	Bond-X	30	Yes	Yes	No
Zymed Laboratories	Mozaic	96	No	No	No

*Usually bar-coded.

plays a major role in this, as it also allows the system to flag missing reagents that are required to execute protocols defined for individual slides. However, it must be stated that the best safeguard of all is proper care and attention to detail by the technician when setting up an immunochemical staining run. When programming runs involve multiple protocols/reagents, technicians must pay particular attention to slide and reagent location, also ensuring that slides are loaded with the specimens facing the correct way up. When using open systems, such as when performing a manual immunochemical staining run, technicians must ensure that all reagents are made correctly in order to achieve optimal and accurate positive staining.

After completion of all staining procedures, desiccation of slides can become a problem with some stainers if the slides are left in the system for extended periods of time. This is avoided in platforms that use a liquid coverslip or that employ regular buffer washes until human intervention.

3. REFERENCES

1. **Grogan TM** (1992) *Am. J. Clin. Pathol.* **98** (Suppl. 1), S35–S38.
2. **Moreau A, Le Neel T, Joubert M, Truchaud A & Laboisse C** (1998) *Clin. Chim. Acta*, **278**, 177–184.
3. **Le Neel T, Moreau A, Laboisse C & Truchaud A** (1998) *Clin. Chim. Acta*, **278**, 185–192.
4. **Biesterfeld S, Kraus HL, Reineke T, Muys L, Mihalcea AM & Rudlowski C** (2003) *Anal. Quant. Cytol. Histol.* **25**, 90–96.

APPENDIX 1

Recipes

Making any % (v/v) solution

For 100 ml final volume of solution, take the % value in ml, subtract this from 100, and add the % value in ml to the answer in ml of ultrapure water*. For example, for 100 ml final volume of a 70% (v/v) ethanol solution, add 70 ml of absolute ethanol to 30 ml of ultrapure water*. A 100% (v/v) solution would therefore indicate that a solution is to be used neat (absolute/undiluted).

*Assume that the solution is ultrapure water-based, unless otherwise specified.

Making any % (w/v) solution

For 100 ml final volume of solution, take the % value in grams, dissolve it in 90 ml of ultrapure water* and make the solution up to a final volume of 100 ml with ultrapure water. For example, for 100 ml final volume of a 1% (w/v) BSA solution, dissolve 1 g of BSA in 90 ml of ultrapure water and make the solution up to a final volume of 100 ml with ultrapure water.

*Assume that the solution is ultrapure water-based, unless otherwise specified.

Making any molar solution

Take the molecular weight of a chemical in grams and dissolve it in 900 ml of ultrapure water*. Make the solution up to a final volume of 1000 ml with ultrapure water.

For example, for a 1 M sodium deoxycholate solution, dissolve 414.55 g of sodium deoxycholate in 1000 ml of ultrapure water. A 4 mM solution would therefore comprise 1.66 g of sodium deoxycholate dissolved in 1000 ml of ultra pure water*.

*Assume that the solution is ultrapure water-based, unless otherwise specified.

Coated slides

Protocol

APES

Equipment and Reagents

- 2% (v/v) 3-aminopropyltriethoxysilane (APES) in absolute acetone
- 95% Acetone
- Glass microscope slides
- Slide rack (load slides into this before commencing the protocol)

Method

1. Wash slides in detergent for 30 min.
2. Wash slides in running tap water for 30 min.
3. Wash slides in two changes of distilled water for 5 min each.
4. Wash slides in two changes of 95% acetone (v/v) for 5 min each.
5. Air dry slides for 10 min.
6. Immerse slides in freshly prepared 2% (v/v) 3-aminopropyltriethoxysilane (APES) in absolute acetone for 5 s.
7. Wash slides in two changes of distilled water for 5 s each.
8. Dry slides in rack overnight at 42°C and store at room temperature.

Protocol

Poly-L-lysine

Equipment and Reagents

- 0.01% (w/v) Poly-L-lysine
- 5% (w/v) Chromic acid (chromium trioxide)
- Glass microscope slides
- Slide rack (load slides into this before commencing the protocol)

Method

1. Wash slides in detergent for 30 min.

2. Wash slides in running tap water for 30 min.
3. Wash slides in two changes of distilled water for 5 min each.
4. Immerse slides in 5% (w/v) chromic acid (chromium trioxide) for 20 min.
5. Wash slides in six changes of distilled water for 5 s each.
6. Immerse slides in 0.01% (w/v) poly-L-lysine for 20 min.
7. Dry slides in rack at 37°C for 24 h and store at room temperature.

Antigen-retrieval solutions

0.1 M EDTA (pH 8)

370 mg EDTA (Sigma product code E-5134)
900 ml Ultrapure water.

Adjust to pH 8.0 using 0.1 M NaOH and make up to 1000 ml with ultrapure water.

0.5 M Tris base buffer (pH 10)

60.57 g Tris base
900 ml Ultrapure water

Adjust to pH 10.0 using 1 M NaOH and make up to 1000 ml with ultrapure water.

For differentiating regressive hematoxylin

1% Acid alcohol

990 ml 70% (v/v) methanol
10 ml Concentrated HCl

For bluing all hematoxylin (in areas where tap water does not have sufficient alkalinity)

Scott's tap water

20 g Sodium bicarbonate
3.5 g Magnesium sulfate
900 ml Ultrapure water

Once the ingredients have dissolved, make up to 1000 ml with ultrapure water and then filter the solution into a separate bottle.

APPENDIX 2
List of suppliers

Abcam plc – www.abcam.com
AbD Serotec – www.serotec.co.uk
Abgene – www.abgene.com
Abgent, Inc. – www.abgent.com
Abnova – www.abnova.com.tw
Academy Bio-Medical Co. – www.academybiomed.com
Active Motif – www.activemotif.com
Aczon Pharma – www.aczonpharma.com
Adobe – www.adobe.com
Advanced Targeting Systems Inc. – www.atsbio.com
Affibody AB – www.affibody.com
Affinity Bioreagents – www.bioreagents.com
AgriSera – www.Agrisera.se
Alba Bioscience – www.albabioscience.com
Aleken Biologicals – www.alekenbio.com
Alexis Corporation – www.alexis-corp.com
AlphaGenix Inc. – www.alphagenix.com
Amersham Biosciences – www.amershambiosciences.com
Anachem Ltd – www.anachem.co.uk
AnaSpec Inc – www.anaspec.com
AngioBio – www.angiobio.com
Anogen – www.anogen.ca
Antibodies for Research Applications BV – www.abforresearch.nl
Antibodies Inc. – www.antibodiesinc.com
Antibody Solutions Inc. – www.antibody.com
Antibody Technology Australia Pty Ltd. – www.antibodytechnology.com
AntibodyShop A/S – www.antibodyshop.com
Aperio Technologies Inc. – www.aperio.com
Appleton Woods Ltd – www.appletonwoods.co.uk
Applied Biological Materials Inc. – www.abmgood.com
Applied Biosystems – www.appliedbiosystems.com

Applied Imaging Corporation – www.aicorp.com
Aprogen Inc. – www.aprogen.com
Ascenion GmbH – www.ascenion.de
Assay Designs – www.assaydesigns.com
ATGen Co. Ltd – www.atgen.co.kr
Aurion – www.aurion.nl
AutoGen Inc. – www.autogen.com
AutoQuant (now part of Media Cybernetics Inc.) – www.mediacy.com
Aves Labs – www.aveslab.com
Aviva Systems Biology Corporation – www.avivasysbio.com
Axon Instruments – www.axon.com
Axxora LLC / Apotech – www.axxora.com

Beckman Coulter Inc. – www.beckman.com
Becton, Dickinson and Company – www.bd.com
Bethyl Laboratories Inc. – www.bethyl.com
Biocare Medical – www.biocare.net
BioChain – www.biochain.com
Biodesign International – www.biodesign.com
BioFX Laboratories – www.biofx.com
Biogenesis Ltd – www.biogenesis.co.uk
BioLegend – www.biolegend.com
BioLogo – www.biologo.de
Biomeda Corporation – www.biomeda.com
BIOMOL International, L.P. – www.biomol.com
Bioptonics – www.bioptonics.com
Bio-Rad Laboratories, Inc. – www.bio-rad.com
Biosense Laboratories AS – www.biosense.com
Biosource Europe S.A. – www.biosource.com
BioVendor – www.biovendor.com
Biovision Inc. – www.biovisionlabs.com
BMA Biomedicals AG – www.bma.ch
BOC Group – www.boc.com
Boston Biochem – www.bostonbiochem.com
British Biocell – www.britishbiocell.co.uk
Brosch Direct Ltd – www.broschdirect.com

Calbiochem – www.calbiochemicom
Cambridge Scientific Products – www.cambridgescientific.com
Capralogics Inc. – www.capralogics.com
Carl Zeiss – www.zeiss.com
Cayman Chemical Company Inc. – www.caymanchem.com

Cell Factors plc – www.cellfactors.com
Cell Sciences Inc. – www.cellsciences.com
Centro Nacional de Biotecnología – www.cnb.uam.es
Chemicon International Inc. – www.chemicon.com
Clonegene LLC – www.clonegene.com
Corning Inc. – www.corning.com
Covalab – www.covalab.com
Covance Research Products Inc. – http://store.crpinc.com/
CRT/Cancer Research Technology Ltd – www.cancertechnology.com/antibodies
CSS-Albachem Ltd – www.albachem.co.uk
Cygnus Technologies Inc. – www.cygnustechnologies.com
Cytomyx Ltd – www.cytomyx.com
CytoStore – www.cytostore.com

DakoCytomation – www.dakocytomation.com
Definiens AG – www.definiens-imaging.com
Delta Biolabs LLC – www.deltabiolabs.com
Detroit R&D Products – www.detroitrandd.com
DevaTal Inc. – www.devatal.com
Diagnostic Biosystems – www.dbiosys.com
Difco Laboratories – www.difco.com
Dionex Corporation – www.dionex.com
DMetrix Inc. – www.dmetrix.com
DuPont – www.dupont.com

East Coast Biologics Inc. – www.eastcoastbio.com
ECM Biosciences, LLC – www.ecmbio.com
Elastin Products Company Inc. – www.elastin.com
Elliot Scientific Ltd – www.elliotscientific.com
EMD Biosciences – www.merckbiosciences.co.uk/home.asp
EnCor Biotechnology Inc. – www.encorbio.com
Epitomics Inc. – www.epitomics.com
Erfa Biotech – www.erfabiotech.com
European Collection of Animal Cell Culture – www.ecacc.org.uk
Everest Biotechnology Ltd – www.everestbiotech.com
Exalpha Biologicals Inc. – www.exalpha.com
Exbio Praha a.s. – www.exbio.cz
Evident Technologies – www.evidenttech.com

FabGennix International Inc. – www.fabgennix.com
Findel Education Ltd – www.fipd.co.uk

Fisher Scientific International – www.fishersci.com
Fitzgerald Industries International, Inc. – www.fitzgerald-fii.com
Fluka – www.sigma-aldrich.com
Fluorochem – www.fluorochem.co.uk
Fusion Antibodies Ltd – www.fusionantibodies.com

Gemac Society – www.gemacbio.com
Genesis Biotech – www.genesisbio.com.tw
GeneTel Laboratories LLC – www.genetel-lab.com
Genetex Inc. – www.genetex.com
Genovac – www.genovac.com
GenWay Biotech Inc. – www.genwaybio.com
GloboZymes – www.globozymes.com
Glycotope GmbH – www.glycotope.com
Good Biotech Corp – www.good-biotech.com/3-2.htm
Goodfellow Cambridge Ltd – www.goodfellow.com
Greiner Bio-One – www.gbo.com

Hamamatsu Photonics – www.hamamatsu.com
Harlan – www.harlan.com
Histopathology Ltd – www.histopat.hu
Hybaid – www.hybaid.com
HyClone Laboratories – www.hyclone.com
HyCult Biotechnology bv – www.hbt.nl
HyTest Ltd – www.hytest.fi

ICN Biomedicals Inc. – www.icnbiomed.com
Imgenex Corporation – www.imgenex.com
Immundiagnostik AG – www.immundiagnostik.com
ImmuneChem Pharmaceuticals – www.immunechem.com
ImmunoDetect Inc. – www.immunodetectantibodies.com
Immunology Consultants Inc. – www.icllab.com
Immuno-Precise Antibodies Ltd. – www.immuno-precise.com
ImmuQuest Ltd – www.immuquest.com
Insight Biotechnology – www.insightbio.com
Invitek GmbH – www.invitek.de
IQ Products – www.iqproducts.nl

Jencons-PLS – www.jencons.co.uk

Kendro Laboratory Products – www.kendro.com
Kodak: Eastman Fine Chemicals – www.eastman.com

LabFrontier Co. Ltd – bio.labfrontier.com/index.asp
Lab-Plant Ltd – www.labplant.com
LAE Bio – www.laebio.com
Lancaster – www.lancastersynthesis.com
Leica – www.leica.com
Leinco Technologies Inc. – www.leinco.com
Life Diagnostics Inc. – www.lifediagnostics.com
Life Technologies Inc. – www.lifetech.com
LOT-Oriel – www.lot-oriel.com

Maine Biotechnology Services Inc. – www.mainebiotechnology.com
Matreya LLC – www.matreya.com
Medignostic – www.medignostic.de
Medix – www.medixbiochemica.com
Merck, Sharp and Dohme – www.msd.com
MetaChem – www.metachem.com
Millipore Corporation – www.millipore.com
Miltenyi Biotec – www.miltenyibiotec.com
ModiQuest B.V. – www.modiquest.com
Mubio Products bv – www.mubio.com
MWG Biotech – www.mwg-biotech.com

Nanoprobes Inc. – www.nanoprobes.com
National Diagnostics – www.nationaldiagnostics.com
Neoclone Biotechnology International – www.neoclone.com
Neomics Co. Ltd – www.neomics.com
Neuromics Antibodies – www.neuromics.com
New England BioLabs Inc. – www.neb.com
Nikon Corporation – www.nikon.com
Novocastra – www.novocastra.co.uk/ce/e/Efullprod.htm
Novus Biologicals Inc. – www.novus-biologicals.com

Olympus Corporation – www.olympus-global.com
Open Biosystems – www.openbiosystems.com
Optivision Ltd – www.optivision.co.uk
Oxford Biomedical Research, Inc – www.oxfordbiomed.com

Peprotech EC Ltd – www.peprotechec.com
Peprotech Inc. – www.peprotech.com
Perbio Science – www.perbio.com
PerkinElmer Inc. – www.perkinelmer.com
Perseus Proteomics Inc. – www.ppmx.com/en/index.html

Pharmacia Biotech Europe – www.biochrom.co.uk
PhosphoSolutions – www.phosphosolutions.com
Photonic Solutions plc – www.psplc.com
PickCell Laboratories B.V. – www.pickcell-b2b.com
ProMab – www.promab.com
Promega Corporation – www.promega.com
ProSci Inc. – www.prosci-inc.com/shop
Protein Biotechnologies Inc. – www.proteinbiotechnologies.com
ProteinOne Inc. – www.proteinone.com
Proteus Biosciences Inc. – www.proteus-biosciences.com

QED Bioscience Inc. – www.qedbio.com
Qiagen N.V. – www.qiagen.com
QuattroMed – www.quattromed.ee

R&D Antibodies Inc. – www.rdabs.com
R&D Systems – www.rndsystems.com
Randox Laboratories Ltd – www.randox.com
ReliaTech GmbH – www.reliatech.de
ReproCELL Inc. – www.reprocell.com/en
Roboscreen – www.roboscreen.com
Roche Diagnostics Corporation – www.roche-applied-science.com
Rockland Inc. – www.rockland-inc.com

Sanbio bv – www.monosan.com
Sanyo Gallenkamp – www.sanyogallenkamp.com
Sarstedt – www.sarstedt.com
Scientifica – www.scientifica.uk.com
Seven Hills Bioreagents – www.sevenhillsbioreagents.com
Shandon Scientific Ltd – www.shandon.com
Sigma-Aldrich – www.sigma-aldrich.com
Signal Antibody Technology Inc. – www.satbio.com
Signature Immunologics – www.immunologics.com
Signet Laboratories Inc. – www.signetlabs.com
Soft Imaging System – www.soft-imaging.net
Sorvall – www.sorvall.com
Southern Biotech – www.southernbiotech.com
Spring Bioscience – www.springbio.com
Statens Serum Institut – www.ssi.dk/antibodies
Stem Cell Sciences – www.stemcellsciencesltd.com
Stratagene Corporation – www.stratagene.com
Strategic Biosolutions – www.strategicbiosolutions.com

Thames Restek – www.thamesrestek.co.uk
Thermo Electron Corporation – www.thermo.com
Thistle Scientific – www.thistlescientific.co.uk
Toxin Technology Inc. – www.toxintechnology.com
Tulip Biolabs – www.tulipbiolabs.com

Vector Laboratories – www.vectorlabs.com
Ventana Medical Systems – www.ventanamed.com
Virogen – www.virogen.com
ViroStat Inc. – www.virostat-inc.com
VWR International Ltd – www.bdh.com

Whatman – www.whatman.com
Wolf Laboratories – www.wolflabs.co.uk

Yamasa Corporation – www.yamasa.com/shindan/english/index.htm
York Glassware Services Ltd – www.ygs.net
Yorkshire Bioscience Ltd – www.york-bio.com

ZeptoMetrix Corporation – www.zeptometrix.com

Index

Bold entries indicate protocols described in the book.